Cases in Systems Analysis and Design:
Best Practices FOURTH EDITION

Cases in Systems Analysis and Design: Best Practices

FOURTH EDITION

Nicholas P. Vitalari

Vice President
CSC Research and Advisory Services

James C. Wetherbe

Professor and Director
MIS Research Center
and Index Research Fellow
Carlson School of Management
University of Minnesota

Federal Express Professor and Director
Federal Express Center for Cycle Time Research
Fogelman College of Business
University of Memphis

West Publishing Company

Minneapolis/St. Paul • New York • Los Angeles • San Francisco

Copyedit: David Dexter
Composition: Parkwood Composition Services, Inc.
Cover image: Warren Gebert

A list of registered trademarks found in this text follows the index.

WEST'S COMMITMENT TO THE ENVIRONMENT
In 1906, West Publishing Company began recycling materials left over from the production of books. This began a tradition of efficient and responsible use of resources. Today, up to 95 percent of our legal books and 70 percent of our college texts are printed on recycled, acid-free stock. West also recycles nearly 22 million pounds of scrap paper annually—the equivalent of 181,717 trees. Since the 1960s, West has devised ways to capture and recycle waste inks, solvents, oils, and vapors created in the printing process. We also recycle plastics of all kinds, wood, glass, corrugated cardboard, and batteries, and have eliminated the use of Styrofoam book packaging. We at West are proud of the longevity and the scope of our commitment to the environment.

Production, Prepress, Printing and Binding by West Publishing Company.

British Library Cataloguing-in-Publication Data. A catalogue record for this book is available from the Britiish Library.

COPYRIGHT © 1979, 1984, 1988   By WEST PUBLISHING COMPANY
COPYRIGHT © 1995   By WEST PUBLISHING COMPANY
                                   610 Opperman Drive
                                   P.O. Box 64526
                                   St. Paul, MN 55164-0526

All rights reserved

Printed in the United States of America

02 01 00 99 98 97 96 95   8 7 6 5 4 3 2 1 0

Library of Congress Cataloging–in–Publication Data

Vitalari, Nicholas P., 1953–
    Cases in systems analysis an design best practices /Vitalari, Nicholas P. Wetherbe, James C. — 4th ed.
      p. cm.
    Rev. ed. of: Cases in systems analysis and design / James C. Wetherbe. 3rd ed. c1988.
    ISBN 0-314-02877-3 (soft)
    1. Management information systems—Cases studies. 2. System analysis—Cases studies. I. Wetherbe, James C. II. Wetherbe, James C. Cases in systems analysis and design. III. Title.
T58.6.V57 1995
658.4'038--dc20   94-43214
                                                  CIP

To
Ann and George
and
Wanda and Bill

Contents

Preface	ix
Chapter 1 — Buffland Companies	**1**
Assignment #1 — Buffland National Bank	1
Assignment #2 — Buffland Power and Light	3
Assignment #3 — Buffland Sales Exchange	4
Chapter 2 — Samson Manufacturing Inc.	**11**
Introduction	11
Executive Meeting	11
First Project Team Meeting	13
Second Project Team Meeting	14
Third Project Team Meeting	19
Executive Presentation	23
Requirements to Complete Case	24
Chapter 3 — New National Bank	**27**
Introduction	27
Executive Meeting	27
First Project Team Meeting	30
Second Project Team Meeting	30
Third Project Team Meeting	37
Executive Presentation	40
Requirements to Complete Case	40
Chapter 4 — Memorial Hospital	**43**
Introduction	43
Executive Meeting	43
First Project Team Meeting	45
Second Project Team Meeting	46
Third Project Team Meeting	51
Executive Presentation	53
Requirements to Complete Case	56

Chapter 5 — Intermountain Distributing Inc.   **57**
Introduction   57
Executive Meeting   57
First Project Team Meeting   59
Second Project Team Meeting   60
Third Project Team Meeting   64
Executive Presentation   67
Requirements to Complete Case   68

Chapter 6 — Medical Supply Inc.   **69**
Introduction   69
Executive Meeting   69
First Project Team Meeting   70
Second Project Team Meeting   72
Third Project Team Meeting   75
Executive Presentation   78
Requirements to Complete Case   79

Preface

Since systems analysis is an applied discipline, the learning experience is compromised when theories and concepts are merely discussed and not applied to industry-oriented problems. This book is designed to address that problem.

Mission of the Book

This book is to be used in conjunction with another text, *Systems Analysis and Design: Best Practices,* West Publishing, written by the authors. However, the cases are written in a manner that allows them to be used with other systems analysis and design texts. Chapter 1, Buffland Companies, provides three increasingly difficult exercises in analysis. The case requires the use of data-flow diagrams. Students are equipped to undertake these cases after completing Chapter 9 of the *Systems Analysis and Design: Best Practices* text.

Chapters 2 through 16 provide four different but equally difficult cases for developing system design specifications. Students are equipped to understand one or more of these cases after completing Chapter 10 of the *Systems Analysis and Design: Best Practices* text. The cases usually require several weeks to solve and can be worked on while students complete the remainder of the *Systems Analysis and Design: Best Practices* text.

Developing Industry-based Cases

The reason for developing industry-based systems design cases for students of systems analysis can be compared to the idea behind "work sets" for accounting students. The primary objective is to provide a comprehensive environment where students can apply their newly learned skills. A problem with developing such cases is that they must be operational within the time constraints of a one- or two-seminar course in systems analysis.

The industry-based cases in this book were constructed in a manner analogous to that used for "connect-the dots" drawings. In "connect-the dots" drawings, dots are strategically located on the lines of a completed drawing; then the lines are removed, leaving only the dots (which can be reconnected to complete the drawing).

To develop industry-based systems design cases, the author has used the following procedure. Completed and operational computer-based information systems for different industries were obtained and analyzed. The most important reporting from the systems, the necessary inputs, and the required processing were identified. The systems were reduced to the equivalent of "dots" by isolating the minimal information needed to understand the requirements of the systems.

Given this definition ("dot equivalent") of a system, the student is asked to fill in the "lines" necessary to precisely specify the system. The systems are of necessity abbreviated, as are accounting "work sets," but the essence of the systems is captured in the cases projects.

Contents of Cases

Analysis Cases: Chapter 1

The systems analysis cases in Chapter 1 provide a brief description of three different businesses and then narratively describe data flaws that the student can transform into data-flow diagrams.

Design Cases: Chapters 2 through 6

The design cases are presented as scenarios in which the student is led through a series of events leading up to the specifications required to complete the case. The student is referred to in second person in the scenario. The standard contents of each case follow:

1. *Industry Discussion.* At the beginning of each case the student is given a brief explanation of the industry from which the case is derived.
2. *Introduction.* In the introduction the organizational context and the student's role as a systems analyst are defined.
3. *Executive Meeting.* The systems analyst attends an executive meeting in which the organizational departments and the various managers are introduced. A major systems development project that is to be addressed with the assistance of a project team is revealed to the systems analyst.
4. *Project Team Meetings.* A series of project team meetings is conducted during which various managers present their information requirements. Definitions of reports, transactions, and processing requirements are stated in narrative form.
5. *Executive Presentation.* The results of the project team meetings are presented to top management. Top management adds a few additional reporting requirements and directs the systems analyst to prepare detailed systems specifications.
6. *Requirements to Complete Case.* The systems analyst is given the final instructions necessary to complete the specifications for the case.

The experience with this approach to systems analysis instruction has been very favorable.

Chapters 2 through 6 include cases in wholesale distribution, manufacturing, banking, and hospitals. Each case is designed as a major semester project requiring twenty-five to forty hours to complete. A case can be worked on by more than one student. If time is a constraint, cases can be abbreviated by eliminating the reports requested during the executive presentation. It should be noted that Chapter 6 is for the advanced systems analysis and design students.

Having a repertoire of industries from which students can select cases is extremely valuable. It allows students not only to apply their technical skills, but also to become familiar with different industries. Accordingly, students can begin to determine areas of career interests and, due to their increased knowledge of an industry or industries, interview for jobs more effectively. That is, students learn about the major information processing in an industry and the vocabulary associated with that information processing. They are, therefore, more comfortable and knowledgeable during job interviews in that industry.

Student responses to the cases have been excellent. The students have been able to complete the cases, have increased their confidence in their technical skills, and frequently comment on the insight they have gained into the rigors required to adequately document systems specifications.

The cases can be used in various ways. Some instructors use manual procedures and paper documentation. With this approach, students can use report layouts, screen layouts, data-dictionary forms and decision table forms provided in the back of the book. At the other extreme, CASE (computer assisted systems engineering), as discussed in the *Systems Analysis and Design: Best Practices* text, may be used to complete the design specifications. Some instructors have even completed the systems by having system analysts classes work with programming classes to develop the software to run the system. The cases are written to be used at either extreme or anywhere in between.

Acknowledgments

In conclusion, we wish to acknowledge the contributions of several individuals. We are especially indebted to V. Thomas Dock, University of Wisconsin, who originally encouraged this book to be written in 1976 and served as consulting editor. The reviewers whose helpful suggestions are reflected throughout the text are: Douglas R. Vogel, University of Arizona; Paul J. Kuzdrall, University of Akron; Jin H. Im, University of New Orleans; Stephen H. Friedman, Queensborough Community College; Vivek Choudhury and Albert L. Lederer, University of Pittsburgh; Richard L. Hartley, Central Michigan University; Robert W. Bretz, Western Kentucky University; Marilyn Bohl, IBM; Thomas I. M. Ho, Purdue University; Roger Hayen, University of Nebraska; Maryam Alavi, University of Maryland; Richard Scamell, University of Houston; Charles Paddock, University of Nevada Las Vegas; Everald E. Mills, Wichita State University; Robert T. Keim, Arizona State University; Kenneth W. Veatch, San Antonio College; Dennis Guster, St. Louis Community College; Gary W. Strong, Drexel University; Warren Briggs, Suffolk University; Sandra Fabyan, Columbus Tech; Rod Conner, University of Richmond; Carl Hossler, Kent State University; Rani Mehta, Computer Applications; M. Bond Wetherbe, Loyola University; Sal March and Dave Naumann, University of Minnesota; Detmar Straub, Georgia State University; Cynthia Beath, Southern Methodist University.

We offer sincere thanks to our patient, tolerant, and forgiving support staff: Donna Leschisin and Wendy Brown for their editing and word-processing efforts.

Finally, we acknowledge the efforts and support of our families—Smoky, Jamie, and Jessie Wetherbe and Martha, Paul, and Joseph Vitalari—who are so supportive in our professional efforts.

As we acknowledge the efforts of all who have contributed, we also assume full responsibility for any inadequacies or discrepancies in the text.

NICHOLAS P. VITALARI
Boston, Massachusetts
JAMES C. WETHERBE
Minneapolis, Minnesota

Buffland Companies

The following systems analysis assignments gradually increase in complexity. The focus of each assignment is developing a data-flow diagram (DFD) to represent the system described. Chapter 9 of the *Systems Analysis and Design: Best Practices* text includes a section on constructing DFDs; please refer to it to assist in completing these assignments.

- *Assignment #1: Buffland National Bank (BNB) Demand-Deposit Accounting System (DDAS)*. The objective of the first assignment is to allow you to work backwards from a Level-0 diagram to a conceptual model of the system. This is an easy assignment, because all that is required is to balance the data flows and identify the appropriate external entities.
- *Assignment #2: Buffland Power and Light (BP&L) Consumer-Billing System (CBS)*. The objective of the second assignment is to allow you to work forwards from a conceptual model of the system to a Level-0 diagram. This is a more difficult assignment.
- *Assignment #3: Buffland Sales Exchange (BSE) Distribution and Cash-Management System (DCMS)*. The objective of the third assignment is to allow you to finish the order-processing and accounts-receivable-receipts part of the DCMS. The conceptual model and the purchase-order and accounts-payable parts of the system are presented.

Assignment #1: Buffland National Bank

Buffland National Bank (BNB) is a small bank in the upper Midwest. The bank provides a variety of services, including loans and checking accounts. It has documented its demand-deposit system as shown in Figure 1-1.

TASK: Develop the conceptual model for the demand-deposit system from the diagram in Figure 1-1 and the following narrative.

Depositor transactions and stop-payment orders are received by a teller. The transaction is confirmed and the depositor provided with a confirmation document. The teller may inquire about a depositor's most recently updated account balance via a computer terminal. Depositor transactions are also entered via the computer terminal.

Figure 1-1 Level-0 Diagram of Demand-Deposit Accounting System

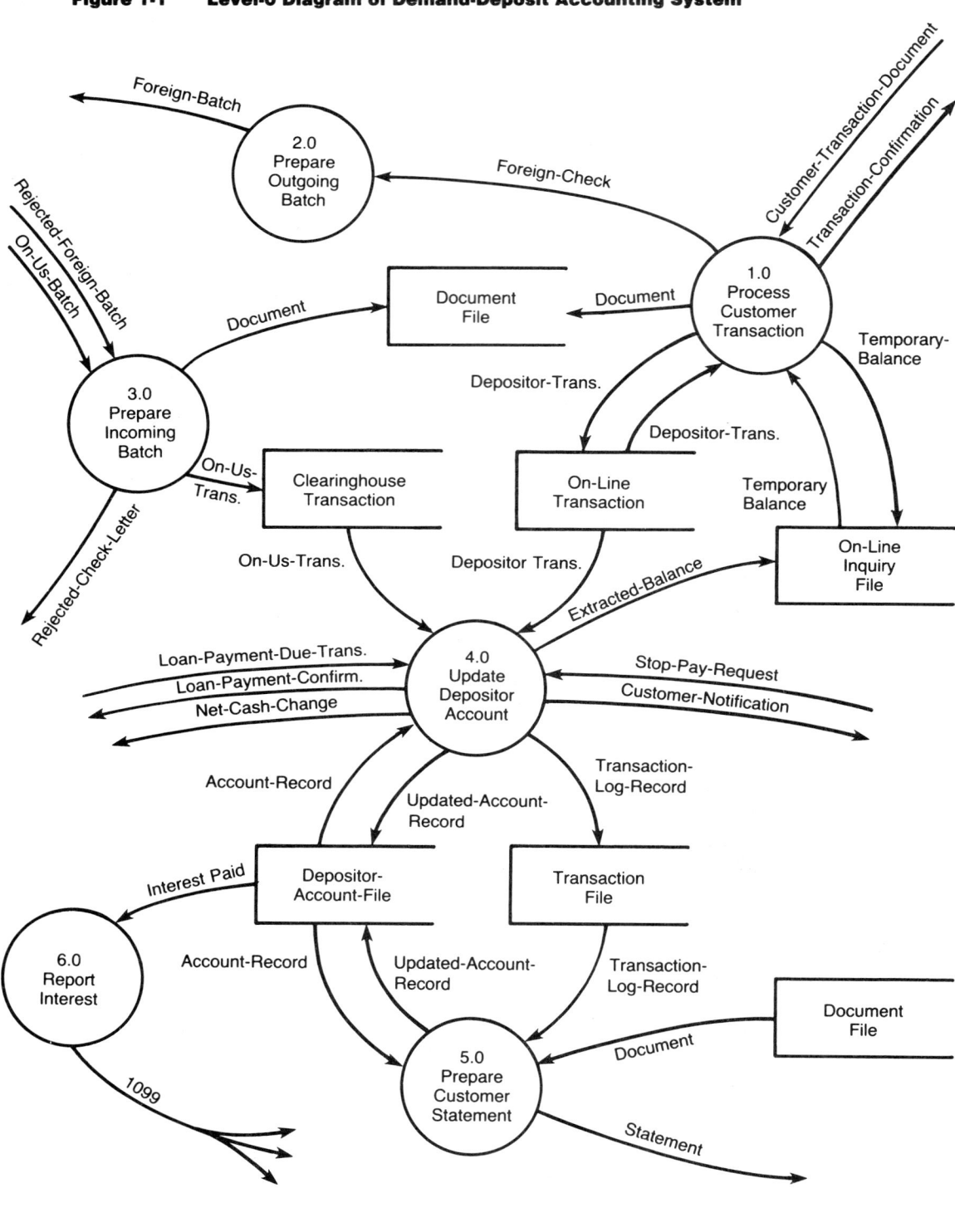

The on-line teller inquiry and update are based upon a temporary on-line database that is recreated each night from the depositor master file after the normal batch processing. Checks drawn on other banks (i.e., foreign banks) are batched together and sent to the clearinghouse. Checks drawn on BNB and received at the teller window are placed in a holding file for batch processing. Rejected foreign checks and checks drawn on BNB but held by other banks, are returned to BNB by the clearinghouse.

Checks drawn on BNB are batched together (both from the clearinghouse and the holding file) with rejected foreign checks. Rejected check letters are sent to the depositor. Checks drawn on BNB are posted to a transaction file.

The depositor's account is updated each night in a batch process. Clearinghouse transactions, on-line transactions, loan payments due, and stop-payment requests are batched together. (Some customers choose to have their loan payments automatically paid from their checking account. The loan payments due are automatically generated by the loan-processing system and paid during the overnight batch processing.)

The update process creates a new temporary on-line file for teller inquiry. Loan-payment confirmation and customer notices of stop-payment requests are sent to the customer. The daily net cash change is reported to the accounting department. Finally, a log of all transactions is created for preparing customer statements at month-end.

Monthly, the accounting department prepares statements from the transaction log and depositor account. Statements and transaction documents (cancelled checks) are sent to the customer. Annually, the accounting department reports the interest paid on each depositor account with Form 1099. Copies of the form are prepared for the IRS, the customer, and the accounting department.

Assignment #2: Buffland Power and Light

Buffland Power and Light (BP&L) is a small electric company in the Midwest. BP&L serves 12,000 rural consumers and has a fairly simple consumer-billing system. A conceptual model indicating the major flows and external entities is shown in Figure 1-2.

TASK: Develop a level-0 diagram for BP&L's consumer-billing system given the conceptual model in Figure 1-2 and the following narrative.

The consumer-services department accepts requests from consumers for connection to the company's electric distribution system. New consumers must pay a connection fee or deposit. The consumer-services clerk creates a *connect order*, which is sent to the meter-reading department for connection. Meter readers make new connects as well as disconnects. The connect order is validated by the meter reader and returned to the billing department. On a monthly basis, meter readers obtain the meter-reading list from the billing department and physically read the electric meters, writing down the new readings on the meter-reading list.

The billing department creates new consumer accounts based upon the connect orders received from meter readers. Every month, meter readings are keyed and validated by the billing department, which updates the electric rates and uses the validated meter readings to generate consumer bills.

Figure 1-2   Conceptual Model of Consumer-Billing System

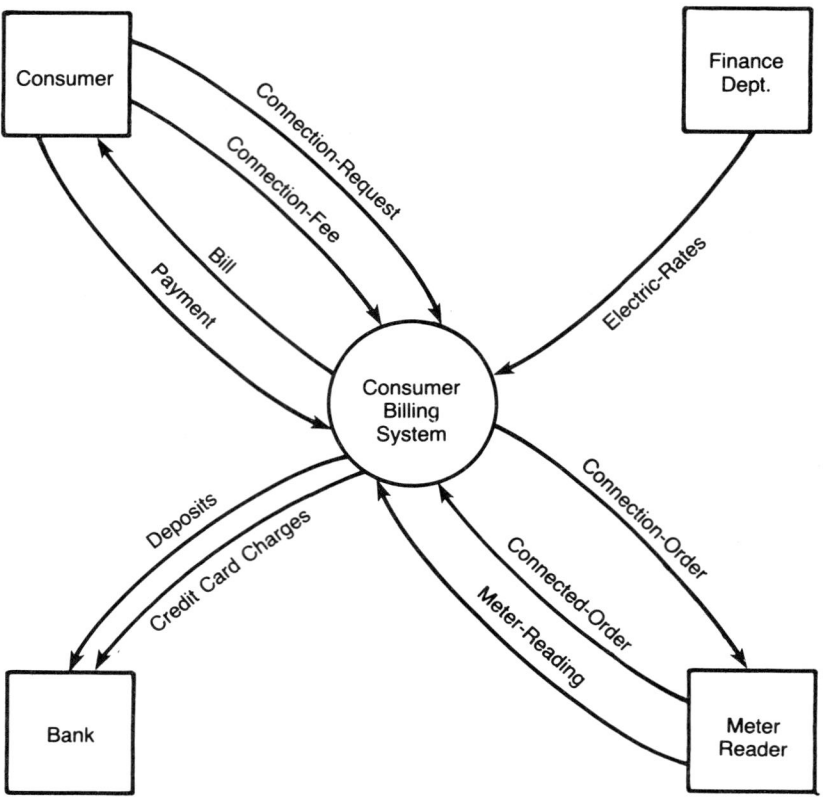

The accounts-receivable department receives payments from consumers and prepares a deposit daily. The consumer's balance due is updated to reflect the payment.

Assignment #3: Buffland Sales Exchange

Buffland Sales Exchange (BSE) is a major retailer of fine fabrics and quality clothing. Although the company occasionally makes bulk sales of its finished goods to a few small retail outlets, the major source of its sales is from its catalog.

The company's distribution and cash-management system is currently being documented with the objective of developing a new system. The systems analysis is incomplete, as shown by the unfinished diagrams in Figures 1-3 through 1-6.

TASK: Develop the missing parts of the company's distribution and cash-management system from the diagrams in the figures and the following narrative.

Customers phone in orders to the company's distribution center. Phoned orders are routed to an order-entry clerk at an interactive terminal. The clerk takes the customer's account number and confirms the customer's billing and shipping address for the order. If the account number isn't known, the clerk conducts a computer search of the customer file, using the customer's name. If this is a new customer, the clerk asks for the billing and shipping address.

Figure 1-3   Conceptual Model of Distribution System

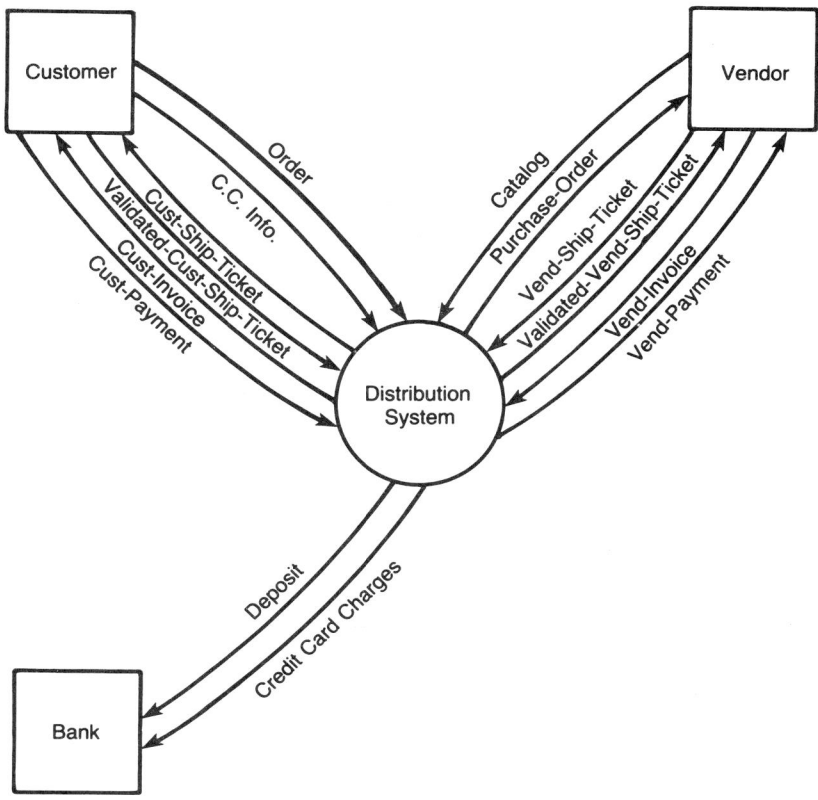

The clerk also asks if the order is to be charged to a major credit card or on open account. Orders charged to a major credit card require that the clerk obtain the credit card number, expiration date, and cardholder's name. The clerk must verify that the credit card number is not on the list of unacceptable accounts provided by the Buffland National Bank. Also, the clerk must place an authorization call if the order is over $50. If the card is not valid or not authorized, the order is cancelled; otherwise, the clerk must record the authorization number provided by the credit card bureau. Orders on account require that the clerk check the customer's account balance and credit limit before taking the order. If the account balance exceeds the credit limit, the clerk must obtain approval from the customer-service manager to accept the order.

Customers order by item number, size, and color, and they give a quantity for each item. The order-entry system locates the item record in the inventory-item file and displays the item description and price. The clerk verifies the item description with the customer and enters the size and color requirements. The system alerts the clerk if the quantity ordered of a particular size and color of an item is out of stock, displaying the quantity that is on order from the factory and the expected time of delivery. If the customer is willing to wait for delivery, the order is placed.

Before ending the phone call, the clerk must tell the customer the sales tax, and the shipping charges, and confirm the total amount ordered. Unshipped orders and backorders are held in a temporary file until shipped.

Figure 1-4   Level-0 Diagram of Distribution System

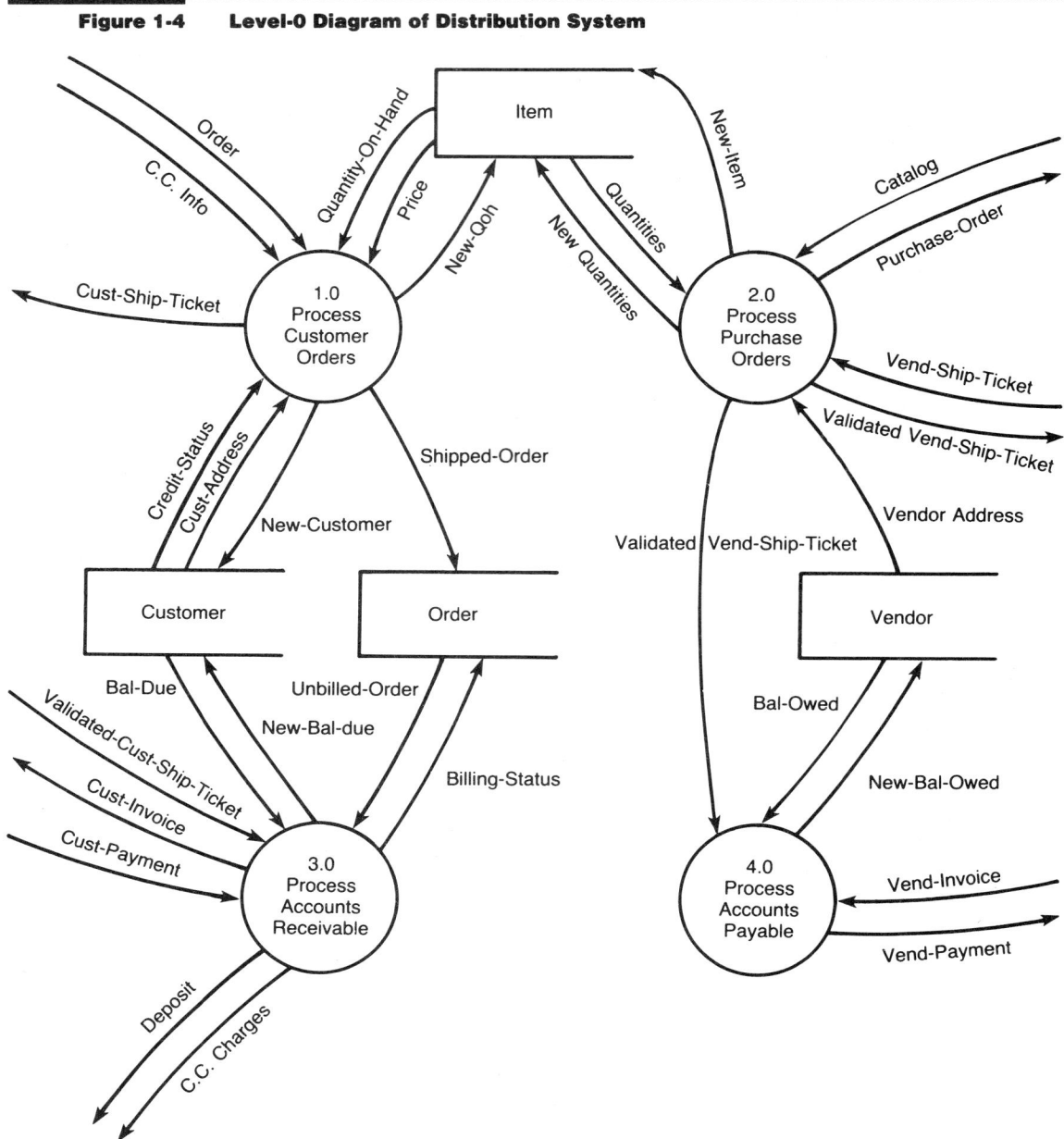

The warehouse clerk periodically processes unshipped orders throughout the day. The clerk prints *picking tickets* from the unshipped-orders file for each order and gives these to the warehouse pickers. The order is filled, and the quantity shipped is marked on the picking ticket by the warehouse pickers. The picking ticket is returned to the warehouse clerk, who verifies the items and quantities shipped and backordered and then updates the unshipped-orders file with the quantities. The clerk also prints a shipping ticket on a serially numbered, multi-part form that shows the quantities shipped and backordered. One copy goes to the customer with the shipment, one copy is filed in the warehouse, and

Figure 1-5   Level 1. Process 2.0—Process Purchase Orders

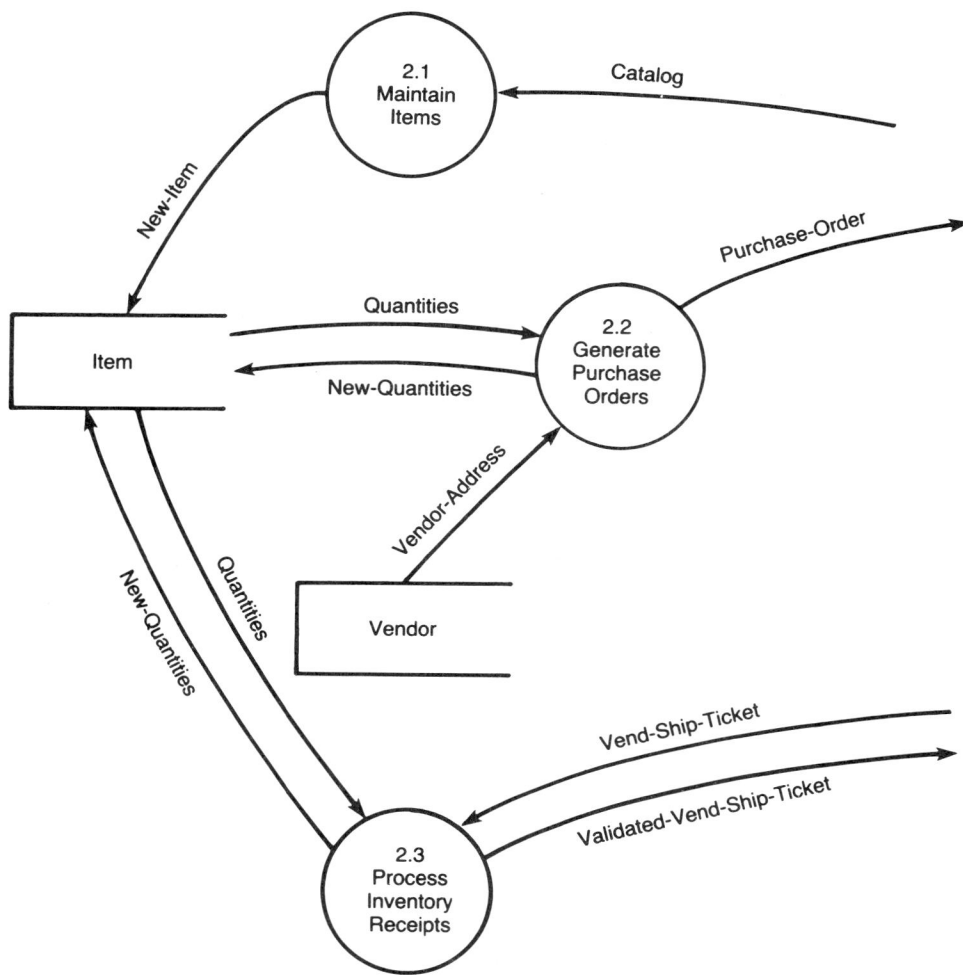

a third copy is sent to accounts receivable. When the shipping ticket is printed, the inventory file is updated to reflect shipment.

The accounts-receivable clerk periodically bills customers for shipped orders. The clerk matches each copy of the shipping ticket with an unbilled order and then generates a customer invoice. The customer account is automatically updated for the new balance due, and the billing status of the unbilled order is updated to reflect the invoice. Unpaid invoice amounts are posted to an accounts-receivable file.

Once per day, the cash-receipts clerk applies credit card and cash payments on account to unpaid invoice amounts. Payments received from customers on *open account* are posted to the customer account and accounts-receivable files. Orders that are completely paid are deleted at this time from the accounts-receivable file. The customer file is updated to reflect the payment. An itemized deposit ticket is automatically prepared during the payment-posting process. *Credit card charges* are extracted from the accounts-receivable file, a list of credit

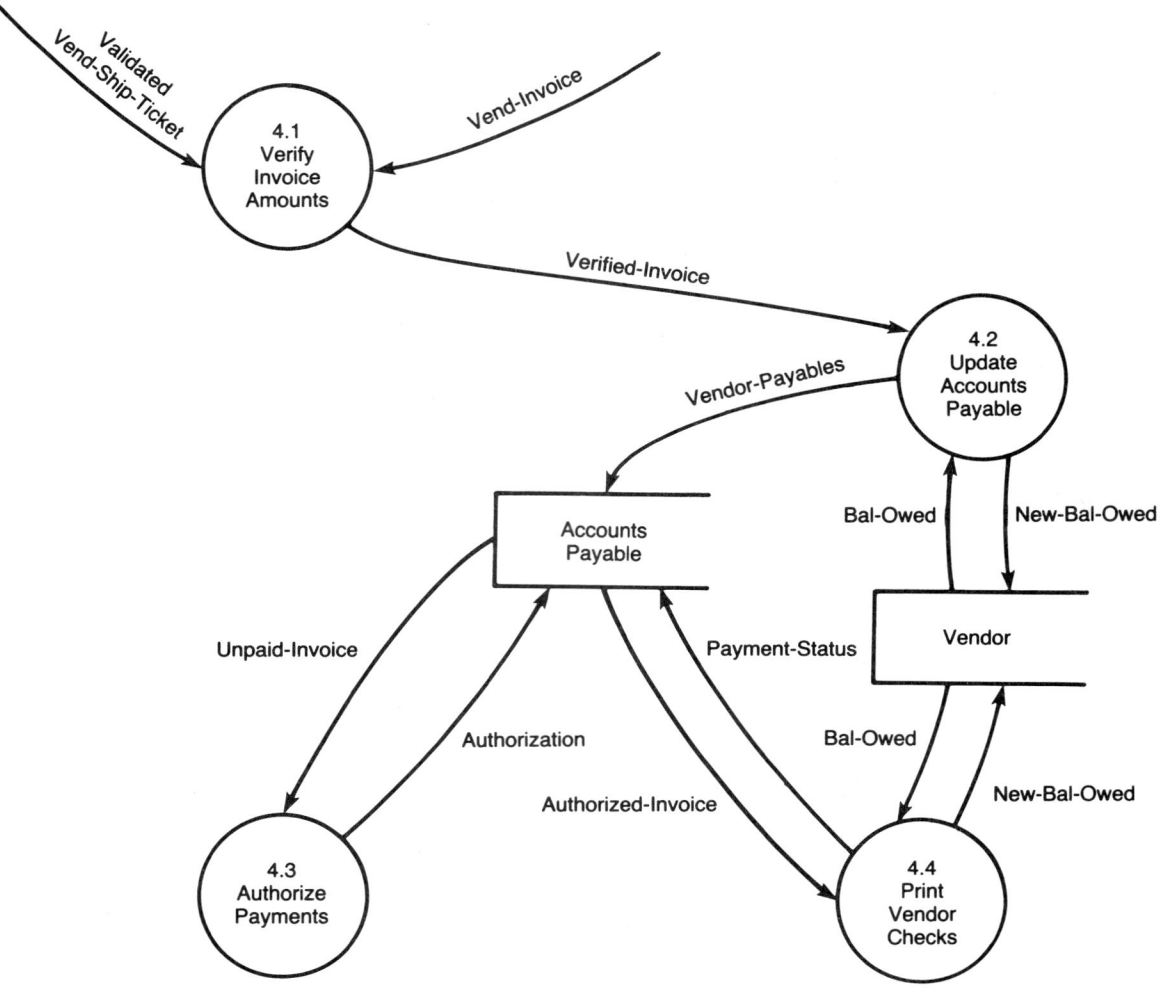

Figure 1-6 Level 1. Process 4.0—Process Accounts Payable

card charges is automatically prepared, and the order is deleted. The deposit ticket and list of credit card charges are sent to the bank with the deposit.

The purchasing clerk periodically updates the inventory-item file with new items from suppliers' catalogs. Each inventory item has associated with it an item number, a description, a quantity on hand, a quantity on order, a reorder point, and a reorder quantity. The purchasing clerk generates purchase orders for each vendor weekly or as necessary. Purchase orders are generated by comparing the quantity on hand plus the quantity on order with the reorder point for each inventory item. If the quantity on hand plus on order is less than the reorder point, the item is reordered using the reorder quantity, and the inventory item file is updated to reflect the new quantity on order. Purchase orders are mailed to vendors.

The warehouse clerk accepts shipments from vendors. Each shipment includes an itemized shipping ticket. The clerk verifies the items and quantities shipped and places the product into inventory. The clerk updates the inventory file to reflect the quantities received—that is, the quantity on order is reduced and the quantity on hand is increased. The vendor shipping ticket is routed to the accounts-payable department.

Accounts payable receives marked-up copies of vendor shipping tickets from the warehouse and invoices from vendors. The accounts-payable clerk matches shipping tickets and invoices. Verified invoices are posted to the accounts-payable file, and the vendor balance owed is updated. The unpaid vendor invoices in the accounts-payable file are authorized for payment by the accounts-payable manager. The payment status is updated by the clerk so that checks may be automatically printed. Vendor invoices that have been authorized for payment are paid weekly or as necessary. The vendor check is printed with data extracted from the vendor and accounts-payable invoice files. Then the vendor balance owed is updated to reflect the payment. Checks are signed by the accounts-payable manager and sent to the vendors.

Chapter 2

Samson Manufacturing Inc.

Manufacturing companies are primarily involved with fabricating, machining, and/or assembling products to be sold to other manufacturers for further processing; to distribution companies for resale; and/or to retail outlets. To be profitable, a manufacturing company must manufacture the appropriate quantities of quality products in a timely and economical manner.

As with all cases in the book, this case is structured to be realistic and representative of the industry for which it is designed. The case is abbreviated so that it can be completed in a short period of time. In particular, the number of reports and terminal displays is reduced. So is the number of data elements. However, the case does include the most important and commonly recognized reporting in the manufacturing industry. ∎

Introduction

You have been hired recently as a systems analyst by Samson Manufacturing Inc. The chief executive officers of Samson have decided to rework their production-information systems, and you have been assigned as the systems analyst to work on the project. Mr. Schroeder, president of Samson, has asked you to attend an executive meeting with the vice-presidents to discuss the new information system.

Executive Meeting

At the executive meeting you are given an abbreviated copy of Samson's organization chart (see Figure 2-1). During the meeting, Mr. Schroeder and Mr. Blair explain that Samson's production activities require information from two basic information systems:

- *Material Requirements Planning*—Provides the information necessary to manufacture the appropriate quantities and qualities of products in an economical

Figure 2-1 Organization Chart for Samson Manufacturing

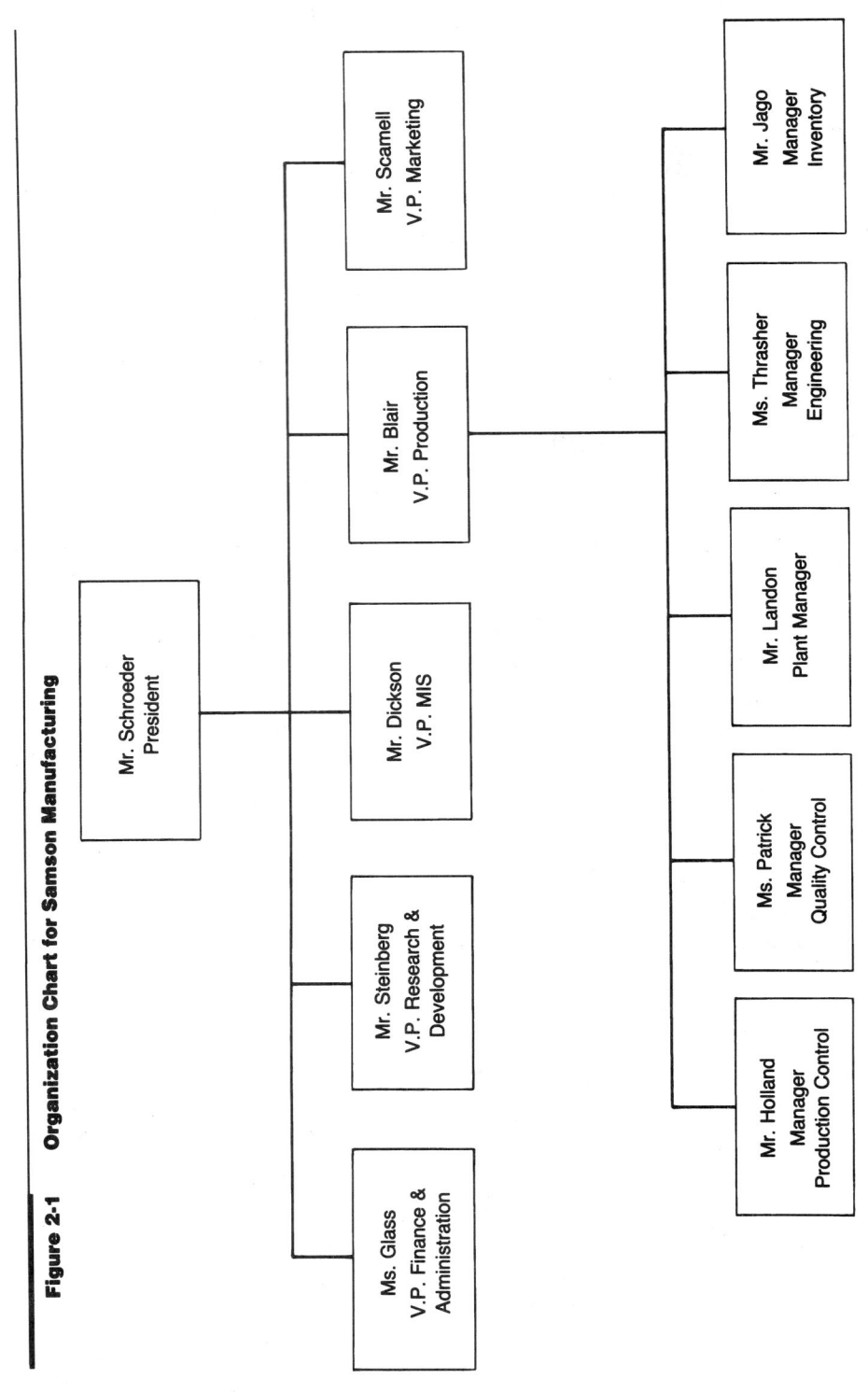

and timely manner. Based upon the number and time when the final products need to be manufactured, the system should determine the material requirements and inform management as to what, how much, and when raw materials, parts, and subassemblies will be required.

- *Production Scheduling and Control*—Provides the information necessary to schedule and control the fabricating, machining, and assembling involved in the manufacturing process.

Mr. Schroeder and Mr. Blair further explain that due to increased volume and complexity in Samson's manufacturing activities, it is becoming increasingly important to have current, accurate information on material requirements and production scheduling and control. Therefore, Samson Manufacturing wants to design a new production-information system that incorporates on-line data entry as necessary.

You suggest that a project team be organized to conduct the necessary analysis and design the new information system. You request that the project team be composed of key management personnel from the production departments affected by the new system.

Mr. Schroeder and Mr. Blair concur with your suggestion. The following personnel are assigned to the project team:

- Mr. Blair, vice-president of production, project sponsor
- Yourself, systems analyst and project coordinator
- Mr. Landon, plant manager
- Mr. Holland, manager of production control
- Ms. Patrick, manager of quality control
- Ms. Thrasher, manager of engineering
- Mr. Jago, manager of inventory

Each team member is charged with determining the information requirements of his or her department. You are assigned overall responsibility for coordinating the team efforts and for developing the systems specifications.

(Note: The requirements for the outputs of the system should be derived using heuristic or prototyping techniques as described in Chapter 10 of *Systems Analysis and Design: Best Practices*. Since this is not practical in a case study, however, outputs are provided for you.)

First Project Team Meeting

You organize a meeting of the project team to initiate the project. You propose that the team address the material-requirements-planning (MRP) system first and then the production-scheduling-and-control (PSC) system. The project team agrees.

You explain that for each system the project team members need to determine the reporting requirements of their departments. You explain that once these have been determined, you will be able to "back into" the inputs and processing needed to generate the reports and terminal displays. You stress that there is cost associated with generating computer-based information, so only the information necessary for operations and decision making should be included in the reports.

You offer to coordinate the work of the team members in determining reporting requirements.

Second Project Team Meeting

After extensive analysis of the production information requirements at Samson, a second project team meeting is called. At this meeting, the project team consolidates the various information requirements for the MRP system into several outputs. The outputs needed are as follows.

Inventory Master Display

Mr. Jago, Mr. Holland, Ms. Patrick, Ms. Thrasher, and Mr. Landon indicate that their respective departments need on-line access to inventory items. Mr. Jago provides a report that includes the necessary content of such access (see Figure 2-2). You indicate the format will have to be changed to fit on a video screen. Mr. Jago says, "No problem; go ahead."

You ask Mr. Jago if PART NUMBER is always six digits. He says it is. Mr. Jago adds that the PART DESCRIPTION can be up to fifteen characters long. He explains the remainder of the report as follows:

1. U/M—the unit of measure for an item. Options are:
   - EA = Each
   - BX = Box
   - DZ = Dozen
   - IN = Inch
   - FT = Foot
   - YD = Yard

2. STOCK LOCATION—the warehouse location of an item. This field always contains two alphabetic characters followed by a digit.

3. ON HAND, ON ORDER, RESERVED, AVAILABLE, and REORDER POINT—numeric fields with a maximum value of 999,999. When printed, these fields are to have commas and zero suppression up to the least significant digit. RESERVED is the number of items reserved for scheduled production. AVAILABLE is computed by subtracting RESERVED from ON HAND. REORDER POINT is the level at which inventory should be ordered (i.e., when AVAILABLE is less than REORDER POINT, ordering should occur).

4. LEAD TIME—the number of working days (i.e., excluding weekends and holidays) required to replenish inventory.

5. PRO 1 PUR 2—a code indicating whether an item is produced by Samson or purchased from a vendor. Allowable code values are:
   - 1 = Produced by Samson
   - 2 = Purchased from vendor

6. DATE LAST ACTIVITY—the last date there was any activity (i.e., reorders, reductions, reservations). It is a calendar date formatted as follows:
   MM-DD-YY

In conclusion, Mr. Jago mentions that the report should be listed in part-number order and should include all parts that Samson stocks. But, he adds, they would need to do ad hoc inquiries using PART NUMBER, PART DESCRIPTION, AND/OR STOCK LOCATION.

Figure 2-2   Inventory Master List

INVENTORY MASTER LIST

DATE 5-07-97

PART NUMBER	PART DESCRIPTION	U/M	STOCK LOCATION	ON HAND	ON ORDER	RESERVED	AVAILABLE	LEAD TIME	REORDER POINT	PRO 1 PUR 2	DATE LAST ACTIVITY
106274	ARM ASSEMBLY	EA	AA1	5,000	10,000	2,000	3,000	15	8,000	1	5-01-97
106278	HUB	EA	AA3	2,000	0	0	2,000	10	1,000	2	4-05-97
107768	½ IN GEAR	EA	AA7	75,000	0	20,000	55,000	12	25,000	2	3-15-97
107774	.042 CRS COIL	EA	AB8	9,000	20,000	5,000	4,000	20	22,000	1	5-06-97
107852	COPPER TUBING	FT	BB1	125,000	0	35,000	95,000	10	35,000	2	2-05-97

Chapter 2   Samson Manufacturing Inc.

Extended Bill of Material

Mr. Landon, Mr. Holland, and Mr. Jago discuss the need for an extended bill-of-material display. They explain that an extended bill of material defines the parts and subassemblies required to manufacture a given assembly, subassembly, or end product, and it is necessary for planning purposes—that is, to determine the components required for a given production order, the quantities required, and the lead time necessary for procuring or manufacturing each component.

Mr. Holland and Mr. Jago present a sample format for the extended bill of material (see Figure 2-3). Mr. Holland explains that there should be a separate display or page for each final assembly, subassembly, or end product. The part number and description of the assembly, subassembly, or end product should be listed under the output heading.

Figure 2-3   Extended Bill of Material

DATE 5-15-97   **EXTENDED BILL OF MATERIAL**

PART 106274 ARM ASSEMBLY

LEVEL 0 1 2 3 4 5 6 7 8 9	PART NUMBER	DESCRIPTION	U/M	QUANTITY REQUIRED	PRO CODE 1 PUR CODE 2	LEAD TIME
0	106274	ARM ASSEMBLY	EA	1,000	1	15
1	576843	STUD 1/2 IN	EA	5,000	2	2
1	432681	NUT 1/2 IN	EA	5,000	2	2
1	468721	ARM	EA	1,000	2	10
2	861231	.042 CRS COIL	EA	1,000	2	18
2	462111	.087 CDS NTRGN	IN	2,460	2	8
1	106228	HUB	EA	5,000	2	6
2	467891	SLEEVE 10 IN	EA	5,000	1	5
2	216871	INSERT	EA	10,000	2	4
3	784234	BEARING 1/4 IN	DZ	8,000	2	10
3	687216	HOUSING 1/4 IN	EA	10,000	1	5
3	644001	LUBRICANT	EA	1	2	5

You ask for an explanation of the LEVEL reporting. Ms. Thrasher explains that a level number is a quantified representation used in engineering drawing. It defines the hierarchy of the various components that must be assembled to complete the product.

Mr. Jago explains that the LEVEL columns on one line may contain a value from 0 (zero) through 9. (The particular value is to line up with the corresponding digit in the LEVEL heading.) The value 0 is assigned to a completed assembly, subassembly, or end product that requires no further action. Parts and subassemblies that are subordinate to a level-0 component are assigned the value 1; parts and subassemblies subordinate to level-1 components are assigned the value 2; and so on. For example, in the sample extended bill of material (see Figure 2-3), three level-3 items (644001, 687216, and 784234) are required to complete the level-2 INSERT (216871). The level-2 items 216871 and 467891 are required to complete the level-1 item HUB (106278). Four level-1 items are required to complete the assembly (level 0).

Mr. Holland indicates that all remaining information on the report except for QUANTITY REQUIRED is defined as discussed for the inventory master list. QUANTITY REQUIRED is the number of the U/M required for each part included in the extended bill of material.

Mr. Holland further stipulates that the extended bill of material should be accessible by part-number.

Material/Parts Requisition

Mr. Landon and Mr. Jago review the need for a material/parts requisition. The requisition needs to be generated in response to a production order for a customer, or an internal production order for the manufacture of a part, assembly, subassembly, or end product. The material/parts-requisition reporting controls material releases; lists all items included in a production order; determines quantity required using extended-bill-of-material information; and assists in the timely movement of material by identifying where to obtain material, and where and when to deliver material.

Mr. Jago presents a sample format for the material/parts requisition (see Figure 2-4). Mr. Jago points out that much of the information included in the report was defined in previous reports. He defines the new information as follows:

1. CUSTOMER NUMBER—six-character field containing two alphabetic characters followed by four digits.
2. PRODUCTION ORDER—five-digit number assigned by the computer when a production order is entered. It is used to uniquely identify a requisition.
3. APPROVED and DATE FILLED—headings to be keyed in by the department that initiates a requisition when the materials are delivered.
4. FILLED BY—heading to be signed by inventory personnel when an order is received, filled, and shipped to the requesting department.
5. DATE—date the requisition was made.
6. DELIVER TO—four-character field in which the first character may be A, B, C, D, E, or F, followed by a dash and two digits ranging from 01 to 55. This value is supplied by the requesting department manager when an order is placed.

Figure 2-4   **Material/Parts Requisition**

SIGNED __DGM__	MATERIAL/PARTS REQUISITION	CUSTOMER NUMBER AB1068
DATE FILLED __5-15-97__		PRODUCTION ORDER
FILLED BY __LSG__		10782

DATE	QUANTITY	PART NUMBER	PART DESCRIPTION	DATE NEEDED	PAGE
5-15-97	1000	106274	ARM ASSEMBLY	5-17-97	1

PART NUMBER	PART DESCRIPTION	STOCK LOCATION	DELIVER TO	U/M	QUANTITY
576843	STUD 1/2 IN	CC1	A–10	EA	5.000
432681	NUT 1/2 IN	DC2	A–10	EA	5.000
468721	ARM	BB7	A–10	EA	1.000
861231	.042 CRS COIL	AC5	A–10	EA	1.000

7. DATE NEEDED—date the material is needed. It is supplied by the requesting department manager when an order is placed.

Ms. Thrasher raises the question of how orders are to be placed. Mr. Landon explains that on-line PCs (personal computers) should be located in each department. The PCs should provide a display that helps department personnel to initiate orders.

You ask if requisitions need to be communicated to the inventory department immediately after production orders are made. Mr. Landon indicates that they do. He asks if a PC could be located in the inventory department to facilitate this. You assure him it can. Mr. Jago cautions that all material requirements for a production order must be reserved when the production order is processed.

Inventory Shortage Report

You ask Mr. Jago how he determines inventory shortages. He explains that he intends to review the inventory master list each week. He mentions that it would be helpful if asterisks or similar flags are used to highlight shortages on the report.

You ask Mr. Jago if reviewing the report once a week is timely enough, since orders are placed daily. Although Mr. Jago agrees it is not, he is reluctant to have a complete inventory master list printed daily. Consequently, you suggest that a daily exception report, listing only inventory items for which there are shortages, be available on-line. Mr. Jago thinks this is an excellent idea and suggests that the report have the same format as the inventory master list. You agree to take care of it.

At the conclusion of this second project meeting, you ask the team members to define the reporting requirements for the production-scheduling-and-control (PSC) system during the next three weeks. Then the next project team meeting will be held.

Third Project Team Meeting

After three weeks, the project team meets again to discuss reporting requirements for the PSC system.

Production Orders

Mr. Landon and Mr. Holland first discuss the production order. They explain that the production order is the key transaction in production scheduling and control. Besides being on-line, a hard copy is needed to travel with an order throughout the production process and document production operation standards.

Mr. Holland explains the necessary contents for the production order (see Figure 2-5). Much of the information needed was defined in the MRP system, he says. He defines the new information as follows:

1. QUANTITY ORDERED—number of parts ordered. It can be up to 999999.
2. ISSUE DATE—date a production order is generated by the computer system.
3. SCHEDULED START DATE, SCHEDULED COMPLETE DATE—dates assigned by the production-control department when a production order is entered through a terminal or PC. The format of these fields is explained in Number 11 of this list.
4. OPERATION SEQUENCE—sequence in which production operations are to be performed. Sequence numbers are four digits, ranging from 0001, in increments of 1, to the total of operations required.
5. WORK CENTER—physical location where work is performed. Work centers are coded as one alphabetic character, a dash, and two digits.
6. OPERATION CODE—specific manufacturing process to be performed during a specific operation. The meaning of the code value can be looked up in a reference manual provided by engineering. The format of the code is two alphabetic characters followed by a hyphen and three digits.
7. SET-UP TIME—time required to prepare a work center to perform an operation. It is expressed in hours or fraction of an hour. The largest set-up time is 15.00 hours.
8. OPERATION STANDARD—standard time required to complete the operation for one unit. The time is expressed in hours, and it may range from .00001 hour to 50.00000 hours. The time required is determined by time measurements conducted by the engineering department.
9. EXTENDED OPERATION TIME—total standard time required to complete an operation. It is computed by multiplying the number of units to be

Figure 2-5   Production Order

PRODUCTION ORDER

PART NUMBER	DESCRIPTION	QUANTITY ORDERED	CUSTOMER NUMBER	ISSUE DATE	SCHEDULED START DATE	SCHEDULED COMPLETION DATE	PRODUCTION ORDER
106274	ARM ASSEMBLY	1000	AB1068	9-10-97	1-171	1-181	10782

OPERATION SEQUENCE	WORK CENTER	OPERATION CODE	SET-UP TIME	OPERATION STANDARD	EXTENDED OPERATION TIME	OPERATION DESCRIPTION	SCHEDULED FINISH
0001	A-10	BB-100	.75	.00120	1.95	RIVET 432614-781162	1-177
0002	A-10	BB-101	.50	.00150	2.00	BOLT 478111-681423	1-177
0003	A-11	CC-110	.25	.02160	21.85	MACHINE 786211-423881	1-179
0004	A-12	DD-567	.30	.00100	1.30	INSERT 644001	1-180
0005	A-12	DD-569	.05	.00160	2.10	INSERT 784234	1-181

produced by OPERATION STANDARD and adding it to SET-UP TIME. This figure may range from .01 to 800.00

10. OPERATION DESCRIPTION—twenty-four character, alphanumeric field used to describe an operation.
11. SCHEDULED FINISH—"manufacturing" days (M-days) scheduled to complete a project. M-days are the work days available during a year (i.e., they exclude weekends and holidays). The first M-day is 1. This number is incremented by 1 for each subsequent day until the end of the year. The format for the M-day is the last digit of the year (e.g., for 1991 the digit 1) followed by a hyphen and the three digits of the M-day. The M-day calendar is set up by the production control manager.

Mr. Holland indicates on-line access to production orders need to be supported. Access keys must include part number, production order, issue, scheduled start, and/or scheduled completion date.

Load Report

Mr. Holland indicates he needs a load report so that he can make scheduling decisions when production orders are entered (to set the start and complete dates). He needs to know on a weekly basis the number of hours or work loaded into each work center and the capacity of the work center. The report should project four weeks into the future.

Mr. Holland provides a format for the load report (see Figure 2-6). You tell him that the format would have to be changed to fit on a video screen. He says that would be fine. He explains the information on the report as follows:

- CAPACITY—total number of hours available at a work center during a given week. This figure is determined by the plant manager and the engineering department. It remains relatively stable.
- LOADED—total number of hours currently scheduled at a work center for a given week. It is computed by adding all production orders scheduled for a given week.
- %—loading level. It is computed by dividing LOADED by CAPACITY.
- BEHIND SCH HRS—amount by which LOADED exceeds CAPACITY during the next four-week period. It is computed by subtracting total LOADED from total CAPACITY. It is printed only when BEHIND SCH HRS is a positive number.
- FUTURE LOAD—sum of all production-order requirements that are scheduled beyond the next four-week period.

Mr. Holland requests that the load report be accessible by work-center sequence.

Quantity Deviation Report

Ms. Patrick indicates the need for a quantity deviation report to report production deviations above or below acceptable ranges. She presents a sample report (see Figure 2-7) and explains the contents as follows:

Figure 2-6   Load Report

LOAD REPORT

DATE 6-1-97

WORK CENTER	BEHIND SCH HRS	1ST WEEK CAPACITY	LOADED	%	2ND WEEK CAPACITY	LOADED	%	3RD WEEK CAPACITY	LOADED	%	4TH WEEK CAPACITY	LOADED	%	FUTURE LOAD
A-01	21.0	80	96.4	121	80	82.1	102	80	86.1	108	80	76.4	95	44.0
A-10		40	32.4	81	40	38.6	96	40	24.2	61	40	12.2	31	
A-11		160	140.5	88	160	155.0	103	160	72.6	45	160	108.7	67	
A-12		240	290.0	121	240	188.6	78	240	104.7	44	240	82.6	34	32.4
B-01		40	52.4	131	40	31.7	79	40	18.4	46	40	8.4	21	
B-12		320	100.0	31	320	152.6	47	320	91.6	29	320	77.5	24	128.9

Cases in Systems Analysis and Design: Best Practices

Figure 2-7   Quantity Deviation Report

DATE 5-05-97			QUANTITY DEVIATION REPORT					
WORK CENTER	PRODUCTION ORDER NBR	PART NUMBER	OPERATION SEQUENCE	OPERATION CODE	QUANTITY ORDERED	PREVIOUS OPN QNTY	QUANTITY COMPLETED	%
A-10	10782	478120	0002	BB-101	1000	1000	975	97.5
A-11	10782	786220	0003	CC-110	1000	975	975	
A-12	10782	786221	0004	DD-567	1000	975	920	94.4

- PRODUCTION ORDER NBR—standard number used for a production order.
- PART NUMBER, OPERATION SEQUENCE, OPERATION CODE, and QUANTITY ORDERED—extracted from the production order under consideration.
- PREVIOUS OPN QNTY—number of items completed from the previous manufacturing operation (i.e., the number of items the previous operation started with, less any losses due to damage or shrinkage).
- QUANTITY COMPLETED—net quantity completed by the current work center (i.e., the number completed, less any losses due to damage or shrinkage).
- %—performance level as computed by dividing QUANTITY COMPLETED by PREVIOUS OPN QNTY. Not to be printed when % = 100.

Mr. Holland asks how the data for PREVIOUS OPN QNTY and QUANTITY COMPLETED are to be collected. You suggest that the information be entered via terminals or PCs, using the production-order display as the order is routed through each work center. Everyone agrees to this idea. You volunteer to redesign the production order to accommodate this change.

Since this is the last reporting requirement identified by the project team, you suggest that the project team finalize the formats of all reports and that you will present them to the executives of Samson. The project team members agree. The meeting is terminated.

Executive Presentation

When the report formats are finalized, you present the proposed reporting system at an executive meeting. Mr. Schroeder and the vice-presidents are impressed with the proposed system. They identify one additional report required by management.

Mr. Schroeder and Mr. Blair discuss the need for an effectiveness report. They explain that the report would facilitate analysis of the productivity of each work center and its supervision. They indicate the report should provide the following information about each work center on a weekly basis:

1. OPERATION CODE for each job performed during the week.
2. QUANTITY ORDERED for each job.
3. EXTENDED OPERATION TIME for QUANTITY ORDERED (including SET-UP TIME).
4. QUANTITY COMPLETED.
5. STANDARD OPERATION TIME EQUIVALENT. This is an adjusted figure to account for cases when QUANTITY COMPLETED is less than QUANTITY ORDERED. The standard operation time equivalent is computed by multiplying QUANTITY COMPLETED by OPERATION STANDARD (see Figure 2-5) and adding SET-UP TIME.
6. REPORTED OPERATION TIME required to complete the operation.
7. EFFECTIVENESS PERCENT, computed by dividing the STANDARD OPERATION TIME EQUIVALENT by the REPORTED OPERATION TIME.

Mr. Landon mentions that there are currently no provisions for capturing reported time on operations. You suggest that the production order could be modified to allow this data to be keyed in. Everyone agrees to this approach.

You volunteer to design the effectiveness display. You ask how the information would be accessed. Mr. Blair indicates it would be accesed by work-center code and that all work centers should be available on-line. Mr. Schroeder says he does not need that much detail. He requests that his report include work-center operations only under the following conditions (HINT: This is a good application for a decision table):

1. If an operation is less than 100 in quantity, include it only if the effectiveness percent is less than 75% or greater than 125% and/or if the quantity completed is less than 75% of the quantity ordered.
2. If an operation is between 100 and 1000 in quantity, include it only if the effectiveness percent is less than 85% or greater than 115% and/or if the quantity completed is less than 85% of the quantity ordered.
3. If an operation is more than 1000 in quantity, include it only if the effectiveness percent is less than 95% or greater than 105% and/or if the quantity completed is less than 95% of the quantity ordered.

After discussing the new report, Mr. Schroeder directs you to complete the overall systems specifications for the new system. The meeting is then concluded.

Requirements to Complete Case

Complete or design all outputs from, and inputs to, the system; create a data dictionary; construct any necessary decision tables and decision trees; identify the necessary files and construct a data model; develop a menu, and construct a DFD. Complete as much as you can of the design specifications after each meeting.

In some instances, the exact input or output format and the validation rules are not told here. You are at liberty to define such issues on a judgmental basis.

There are many ways this case can be expanded into a major time-consuming effort. To avoid this, stay within the requirements of the case.

The best way to begin is to take the first report, complete the output layout, and make initial entries in the data dictionary for all data elements in that output. Do the same thing with the second output, the third, and so forth. If you have a CASE tool at your disposal, the output layout can be linked directly to the data dictionary. This will also help ensure consistency in your data definitions.

New National Bank

3

Banks are engaged primarily in providing services associated with checking accounts, savings accounts, and installment, commercial, and mortgage loans. The checking and savings accounts bring money into the bank from which the bank generates revenue by investing in loans. To be profitable, a bank's investment revenue must exceed its interest payments to savings accounts and its operation expenses (e.g., salaries, materials).

As with all cases in the book, this case is structured to be realistic and representative of the industry for which it is designed. The case is abbreviated so that it can be completed in a short period of time. In particular, the number of reports and terminal displays is reduced. So is the number of data elements. However, the case does include the most important and commonly recognized reporting in the banking industry. ∎

Introduction

You have been hired recently as a systems analyst by New National Bank. The chief executive officers of New National have decided to develop a new information system to process DDA (demand-deposit accounting, or checking accounts), savings accounts, and loans. You have been assigned as the systems analyst to participate in the project. Ms. Karpe, President of New National Bank, has asked you to attend an executive meeting with the vice-presidents to discuss the new information system.

Executive Meeting

At the executive meeting you are presented a copy of New National's organization chart (see Figure 3-1). Ms. Karpe explains that Mr. Price, vice-president of operations, and his staff are primarily responsible for DDAs and savings accounts. Mr. Pender, vice-president of lending, and his staff are primarily responsible for making decisions on loans and administering loans. Mr. Adamson, controller,

Figure 3-1   Organization Chart for New National Bank

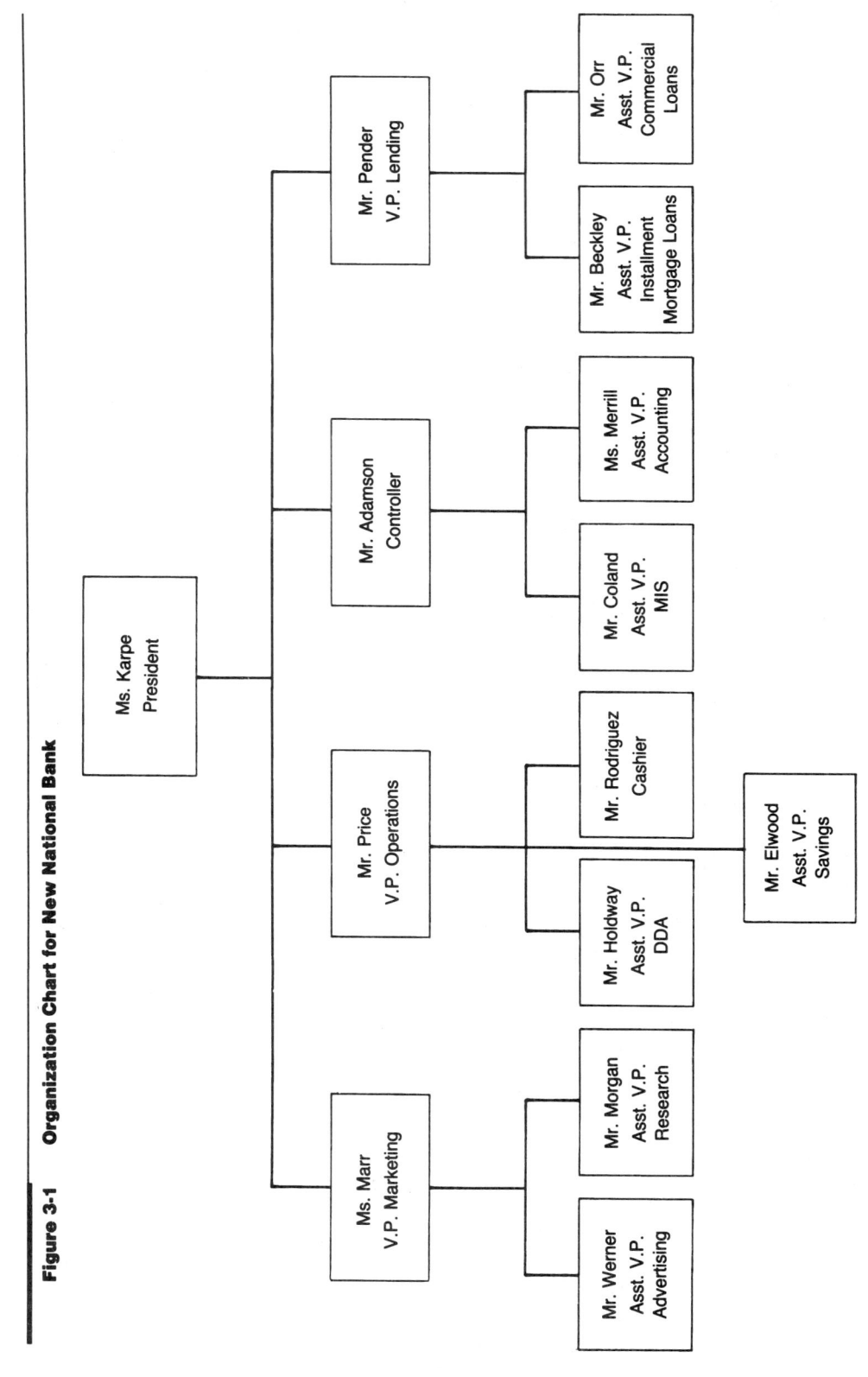

and his staff take care of accounting and computer processing. Ms. Marr, vice-president of marketing, and her staff are responsible for researching and implementing programs to enhance the public image and market competitiveness of New National.

Ms. Marr explains that the research efforts of Mr. Morgan indicate a strong interest among existing and potential customers of New National for the bank to service customers by providing consolidated statements of banking activities. Under this approach, each customer would receive one statement reflecting the status of his or her DDAs, savings accounts, and loans each month.

Ms. Marr further points out that consolidating information about customers will also help the bank. A complete profile of a customer's various accounts would be useful for marketing promotions.

Mr. Pender indicates that consolidating information about customers will also help in making prompt decisions on loans. Currently, the bank takes up to two days to review all the accounts of a customer (i.e., DDA, savings, and commercial, installment, and mortgage loans) before making a decision on a new loan. Since these are reported in different places, this activity is time-consuming, expensive, and results in poor service to the customer.

You ask what the current state of computer-based information systems is at New National. Mr. Adamson explains that New National is in the same position as many banks. It has all major applications on the computer. It uses magnetic-ink character recognition (MICR) equipment in processing checks and savings accounts.[1] All loan transactions are handled by MICR turnaround documents given to customers when loans are originated. The customers then submit monthly payments with the corresponding MICR turnaround documents.

Mr. Adamson indicates that the problem with the information systems at New National is that there is limited capability for integration of information. When DDA was computerized, the system was designed to meet the needs of the DDA department. When savings accounts were computerized, the system was designed to meet the needs of the savings accounts department. The DDA and savings accounts systems use different account numbers, so there is no convenient way to determine whether or not a customer who has a checking account with New National also has a savings account.

Subsequent design and implementation of systems for commercial, installment, and mortgage loans proceeded in the same manner as the DDA and savings-accounts systems. Consequently, New National has customers with multiple checking, savings, and loan accounts, but it has no way to efficiently consolidate the information about each customer. One customer with multiple accounts looks like "multiple customers" to the bank. This results in redundant data capture, processing, storage, and reporting. New National often sends several statements (in different envelopes) to a customer.

Ms. Karpe explains that the inability to consolidate information for decision making and for the convenience of customers, and the obvious inefficiencies associated with processing redundancies, have prompted New National to undergo a major redesign of its banking applications to form an integrated information system. She further explains that she wants to move to on-line processing wherever possible.

[1.] If you are not familiar with MICR technology, you are encouraged to visit a bank that uses MICR.

You suggest that a project team be organized to conduct the necessary analysis and to design the new information system. You request that the project team be composed of high-level staff representatives from all major functions affected by the new system.

Ms. Karpe concurs with your suggestion and request. The following individuals are assigned to the project team:

- Ms. Marr, vice-president of marketing and project sponsor
- Yourself, systems analyst and project coordinator
- Mr. Holdway, assistant vice-president of DDA
- Mr. Elwood, assistant vice-president of savings
- Mr. Beckley, assistant vice-president of installment and mortgage loans
- Mr. Orr, assistant vice-president of commercial loans
- Mr. Rodriguez, cashier
- Ms. Merrill, assistant vice-president of accounting

Each team member is charged with determining the processing and information requirements of his or her organizational area. You are assigned overall responsibility for coordinating the team efforts and for developing the systems specifications.

(Note: The requirements for the outputs from the system should be derived using heuristic or prototyping techniques as described in Chapter 10 of *Systems Analysis and Design: Best Practices*. Since this is not practical in a case study, however, outputs are provided for you.)

First Project Team Meeting

You organize a meeting of the project team to initiate the project. You ask each team member to determine the information requirements of his or her area, explaining that once these have been determined, you will be able to "back into" the inputs and processing needed to generate the reports. You also explain that there is cost associated with generating computer-based information, so only information necessary for operations and decision making should be included in the reports.

You indicate that you will coordinate the team efforts to ensure integration of the information requirements. For example, basic customer demographic data (name, address, phone number, and so on) should be processed and stored only once per customer, irrespective of the number of accounts the customer has.

Second Project Team Meeting

After extensive analysis of the information requirements at New National, a second project team meeting is called. At his meeting, the project team consolidates the various information requirements into several reports.

Consolidated Customer Statement

The first issue to be addressed is the concept of a consolidated customer statement. After considerable discussion, it is agreed that a uniform account structure is required to integrate the information from a customer's various accounts. The account structure agreed upon consists of a nine-digit master account number followed by a two-digit suffix (see Figure 3-2). Each customer will have one unique master account number. Suffixes will be used to uniquely identify different accounts of a particular customer.

After the account number structure is agreed upon, Mr. Holdway, Mr. Elwood, and Mr. Beckley present their proposal for a consolidated customer statement (see Figure 3-3).

You ask for definitions of the contents of the consolidated statement. Mr. Holdway explains that the customer name and address are to be printed in the box so that when the statement is folded, the customer name and address can be seen through an envelope window. He points out that the customer number is the master account number defined earlier. Since a statement may require more than one page, page numbers should be provided.

Mr. Holdway explains the remaining contents of the statement as follows:

1. ACCOUNT NUMBER—name of the category of the account, followed by the account suffix. The account category is indicated by the value of the account suffix as follows:
   01–25 = CHECKING
   26–40 = SAVINGS
   41–60 = INST. LN. (INSTALLMENT LOAN)
   61–80 = COMM. LN. (COMMERCIAL LOAN)
   81–99 = MORT. LN. (MORTGAGE LOAN)

Figure 3-2   Account Number Structure

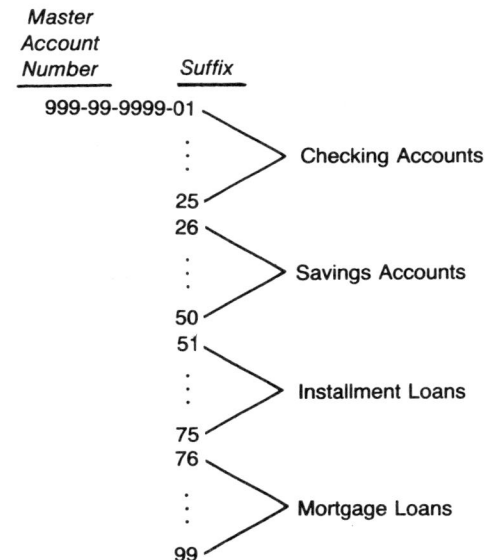

Figure 3-3 Consolidated Customer Statement

NEW NATIONAL BANK
HOUSTON, TEXAS

MR. JOHN DOE
1218 BURNWOOD
HOUSTON, TEXAS, 77073

CUSTOMER NO.	PAGE NO.
124-35-6789	1
STATEMENT PERIOD	
5-29-97 TO 06-01-97	

ACCOUNT NUMBER	BALANCE FORWARD	NUMBER OF CREDITS	TOTAL CREDITS	NUMBER OF DEBITS	TOTAL DEBITS	FEES AND EARNINGS	CLOSING BALANCE
Checking 01	432 75	2	225 60	4	375 00	2 00−	281 35
Checking 02	88 20	1	1,910 00	2	957 20	2 00−	1,039 00
Savings 26	1,670 00					8 35+	1,678 35
Inst. Ln. 41	580 30	1	125 10			5 80	461 00
Inst. Ln. 42	850 60	1	230 25			8 50	628 85
Comm. Ln. 61	7,725 30	2	1,605 60			64 37	6,184 07
Mort. Ln. 81	58,432 20	1	710 20			585 20	58,307 20

TRANSACTIONS

ACCOUNT NUMBER	AMOUNT	AMOUNT	AMOUNT	DATE
Checking 01	100 00 DP	150 30 CK	50 00 CK	06-10
	10 15 CK	164 35 CK	125 60 DP	06-21
Checking 02	180 20 CK			
	777 00 CK	1,910 00 DP		06-05
Inst. Ln. 41	580 30 PM			06-18
Inst. Ln. 42	230 25 PM			06-28
Comm. Ln. 61	1,000 00 PM			06-27
Mort. Ln. 81	710 20 PM	605 60 PM		06-28

CODE EXPLANATION—SEE OTHER SIDE

2. BALANCE FORWARD—balance brought forward from the previous month. It should never exceed 9,999,999.
3. NUMBER OF CREDITS—number of credit transactions during the month. It should never exceed 1000.
4. TOTAL CREDITS—cash amount of total credits for a month. It should never exceed 9,999,999.
5. NUMBER OF DEBITS—number of debit transactions during the month. It should never exceed 1000.
6. TOTAL DEBITS—cash amount of total debits for a month. It should never exceed 9,999,999.
7. FEES AND EARNINGS—expenses and/or revenue of an account during a month. Possible FEES and EARNINGS follow:
    CHECKING—$2 service fee unless customer has $200 or more in checking, in which case there is no fee. The customer pays an additional $2 fee if more than twenty transactions (i.e., debits and/or credits) occur on a checking account in a month. The additional fee does not apply to accounts with $200 or more in checking. Fees are always followed by a minus sign.
    SAVINGS—a savings account earns the interest rate assigned to the customer's account. Interest earnings are prorated on a monthly basis. The customer is charged a $2 fee if a savings account has more than five transactions in a month. Earnings are always followed by a plus sign.
    INST. LN., COMM. LN., MORT. LN.—fee (the interest charge) is prorated on a monthly basis and is always followed by a minus sign.
8. CLOSING BALANCE—computed by adding TOTAL CREDITS and EARNINGS to; and subtracting TOTAL DEBITS and FEES from, BALANCE FORWARD.
9. TRANSACTIONS—itemized details for each transaction: ACCOUNT NUMBER, AMOUNT, and DATE (DATE is formatted mm-dd). The form allows up to four transactions to be listed on the same line. If more than four transactions occur on one day, the next line is used to display additional transaction(s). Each transaction AMOUNT is followed by a transaction code. The possible code values and their meanings are:
    DP = Deposit to a savings or checking account
    CK = Check written against a checking account
    WD = Withdrawal from a savings account
    PM = Payment on a loan
    LN = Additional principal added to an existing loan or initial principal for a new loan

The project team members are satisfied with the format of the consolidated customer statement. It is approved.

Mr. Holdway indicates that New National also needs on-line access to customers and customer-account information. You mention that since a screen has less space than the customer report, the format might have to be changed. Mr. Holdway agrees to this. He also stipulates that the customer accounts should be accessible by account number or customer name.

MICR-Entry Report

Mr. Rodriguez discusses the need for an on-line MICR-entry report. He explains that transactions are sent in batches to the information systems department. He

Figure 3-4   MICR-Entry Report

	MICR-ENTRY REPORT			DATE 7-05-97	
BATCH NO. 102					
ACCOUNT NO.	AMOUNT	ACCOUNT NO.	AMOUNT	ACCOUNT NO.	AMOUNT
586-24-7123-01	8.93	782-42-4687-01	1,689.40	784-36-4777-01	78.40
478-96-8742-01	486.47	432-68-7844-20	50.00	742-78-4928-05	32.00
681-42-7842-03	620.00	333-78-4932-21	872.00	784-36-4777-01	781.20

HEADER TOTAL	6,889.80	HEADER ITEM COUNT	48
ACTUAL TOTAL	6,889.80	ACTUAL ITEM COUNT	48
DIFFERENCE	.00	DIFFERENCE	0

says that, for control purposes, each batch of transactions has a header record that indicates the number of transactions and the total dollar value for the batch. As each batch of transactions is run through the MICR sorter and entered into the computer, the cashier's office needs a list of all items and control totals to check against the batch totals.

He proposes a format for the MICR-entry report (see Figure 3-4). Since the contents of the report are straightforward, few questions are asked. Mr. Rodriguez explains that the transactions are entered into the computer on the first pass of the sorter. Therefore, the transactions will not be sorted until after the MICR-entry report has been generated. The transactions are to be listed sequentially, "three-up" across the screen, and should be accessible by batch number and date. The ACTUAL TOTAL and ACTUAL ITEM COUNT are computed as the transactions are processed by the computer. They are compared to the HEADER TOTAL and HEADER ITEM COUNT to compute each DIFFERENCE.

Posting Journal

Mr. Holdway says that he, as well as the other managers, needs an on-line daily posting journal that reflects all transactions performed against customer accounts. He proposes a format for the report (see Figure 3-5). You indicate it might have

Figure 3-5 Posting Journal

DATE 7-05-97				POSTING JOURNAL						PAGE 10
CUSTOMER	ACCT	OLD BALANCE	DATE LAST TRAN	AMOUNT	AMOUNT	AMOUNT	AMOUNT	AMOUNT	AMOUNT	NEW BALANCE
772-43-6871 BILL GLASS	CK01	362.30	6-10-97	32.00CR	28.00CR	10.00DB				412.30
781-24-6371 CHERYL GLASS	CK01	1,781.20	06-01-97	100.00CR 1,000.00DB	22.00DB 50.00CR	20.00CR	30.00CR	2,000.00CR		2,959.20
532-46-8711	SA26	5,742.80	05-05-97	1,000.00CR	18.00DB					6,742.80
LINDA GLASS	IL42	1,642.80	06-05-97	42.50CR						1,600.30
711-41-6111	ML81	38,572.31	06-05-97	420.00CR						38,152.31
WANDA GLASS	CL62	10,784.70	06-06-97	500.00CR 720.00CR	30.00CR 28.00CR	78.00CR 92.00CR	100.00CR	50.00CR		9,186.70

Chapter 3   New National Bank

to be reformatted to fit on a video screen. He says that will be fine. The posting journal should list all customer accounts that had activity the previous day and should be accessible by account number and customer name. Note that under the heading CUSTOMER, the customer number is printed first, and the corresponding name is printed right below it. Mr. Holdway points out that most of the report contents are already defined in other reports. He defines new information (and previously defined information presented in a different form) as follows:

1. DATE—date of the report
2. ACCT—classification and number of the account. To save space, this information is abbreviated as follows:
   CK = Checking account
   SA = Savings account
   IL = Installment loan
   CL = Commercial loan
   ML = Mortgage loan
3. OLD BALANCE—customer balance prior to transaction posting. In the case of checking and savings accounts, the balance is a customer asset. In the case of loans, the balance is the principal remaining on the loan.
4. DATE LAST TRAN—last date there was any transaction activity for the account. It is updated whenever a transaction occurs.
5. AMOUNT—detailed listing of all debits and credits to the account. The symbol DB (for debit) or CR (for credit) follows each AMOUNT value.
6. NEW BALANCE—new balance for the account value after all transactions have been posted. It is computed by adding all credits to, and subtracting all debits from, the OLD BALANCE.

Trial Balance

Next Ms. Merrill indicates that she needs trial balance figures after all transactions have been posted. She requests a trial balance report, listing all categories of accounts (i.e., DDA, savings, installment loans, commercial loans, and mortgage loans). It should show the balance for each account within a category (i.e., DDA01, DDA02, . . ., DDA25).

You ask if the balance should include the balances for all accounts or just those experiencing transactions the previous day. She answers that it should be the balance of all accounts. For example:

	DDA01	$ 2,432,628.03
	02	897,432.20
		1,785,672,30
		—
		—
		—
	25	786,432.30
TOTAL		$32,468,721.60

You indicate that the trial balance could be accomplished during processing for the posting journal report. The program could sequentially pass all accounts, posting to accounts having transactions, and picking up balances for all accounts. The balances could then be accessed on-line.

Everyone agrees that this would be an efficient way to accomplish reporting for the posting journal and the trial balance. You offer to design a format for the trial balance.

This concludes the second project team meeting. The project team agrees to meet in three weeks to continue defining reporting requirements.

Third Project Team Meeting

Customer Master Display

At the third project meeting, the first report discussed is a customer master terminal display. All project team members had expressed a need for a computer terminal display whereby accurate information about a given customer could be obtained upon request. After considerable discussion, the project team agrees on the contents for the display (see Figure 3-6). Much of the customer master information is already defined in previous reports. The new information is defined as follows:

1. H-PHONE, O-PHONE—home and office phone numbers, respectively.
2. HOME—current housing of the customer. Possible values are:
   1 = Owns; 2 = Rents.
3. SEX—coded as follows:
   1 = Male
   2 = Female
   3 = Couple
4. MARITAL—coded as follows:
   1 = Single
   2 = Married
   3 = Divorced
   4 = Widowed
5. CHILD—number of children
6. EMPLOYER—name of the employer of the first name given in the customer name (in Figure 3-6, of Joe).
7. EMPLOYER-S—for married customers, the employer of the spouse (in Figure 3-6, of Gina).
8. YRS-EMP, YRS-EMP-S—number of years employed by current employer (S designates spouse and is used only when applicable).
9. INCOME, INCOME-S—income (S designates spouse and is only used when applicable).
10. DDA—checking account numbers
11. BALANCE—current balance in a checking account.
12. YR—year an account was opened.

Figure 3-6 Customer Master Display

CUSTOMER MASTER

CUSTOMER NO. 585-12-7864

NAME AND ADDRESS   H-PHONE   HOME   MARITAL   EMPLOYER
                            712-6841   1   2   PIZZA HUT
MR. JOE AND MS. GINA CHILDS
1218 MAIN   O-PHONE   SEX   CHILD   EMPLOYER-S
HOUSTON, TEXAS 77073   744-3177   3   2   EXXON

DDA	BALANCE	YR	NSF	SA	BALANCE	YR	INT-R	YRS-EMP	
01	432.20	81	0		1,562.50	82	.065	YRS-EMP-S	3
02	25.10	82	1	26					3

                                                              INCOME   16,000
                                                              INCOME-S 16,000

LOANS	O-BALANCE	C-BALANCE	PAYMENT	INT-RATE	INT-PD	PMTS	LT-PMTS	DT-LAST-PMT	PMT-DUE
41	785.00	582.75	35.00	.120	85.62	12	0	05-03-97	06-05-97
42	2,200.00	1,871.50	136.00	.120	141.70	18	1	05-03-97	06-05-97
81	39,500.00	38,500.00	398.00	.095	2,658.30	360	0	05-03-97	06-05-97

13. NSF—number of checks written when there were not sufficient funds.
14. SA—savings account numbers.
15. INT-R—interest rate (entered through terminal by the savings department when an account is opened).
16. LOANS—loan numbers.
17. O-BALANCE, C-BALANCE, (LOAN)—O is the original principal borrowed; C is the principal remaining.
18. PAYMENT—monthly payment (entered by a loan officer when loan is originated).
19. INT-RATE—interest rate on a loan (entered through a terminal by a loan officer when loan is originated).
20. INT-PD—interest paid on the loan since loan origination (computed after each payment).
21. PMTS—number of payments that have been made.
22. LT-PMTS—number of payments that have been late (fifteen days later than the due date).
23. DT-LAST-PMT—date the last payment was made.
24. PMT-DUE—date the next payment is due (advanced one month each time a payment is made).

The project team members want to be able to access customer master information by either customer number or customer name. You agree to develop a screen format for the display.

Control Reports

Mr. Holdway explains that he needs a daily report indicating all checking accounts actioned without sufficient funds (NSF accounts). He indicates that if a customer's debit(s) exceeds his or her balance, the debit(s) is not to be posted. An exception report listing customers with NSF accounts is required.

You tell him that you are glad he mentioned this requirement. It means that all transactions have to be sorted so that credit transactions are posted before debit transactions. Otherwise, a customer may be charged erroneously for an NSF account. Mr. Holdway agrees with your observation.

You propose that the exception report be prepared as a by-product of creating the posting journal. You ask Mr. Holdway what the report should contain. He requests that it contain customer name, address, office and home phone numbers, old balance, and the transactions posted. You agree to design a screen format.

Next Mr. Beckley and Mr. Orr indicate they need a monthly exception report of all customers whose loans are more than thirty days delinquent in payments. You indicate that since date-of-last-payment and next-payment-due-date are already available, a simple comparison can be used to detect delinquent accounts. A program could be run at the end of the month to check the dates and report all exceptions.

Mr. Beckley and Mr. Orr agree. They ask if installment, commercial, and mortgage-loan categories could be separated on the report. You say this could be accomplished by sorting the data before printing. You ask what information the report should contain. They indicate they need customer number and name, address, office and home phone numbers, current balance, date-of-last-payment, and next-payment-due-date. You agree to design a screen format.

Since this is the final reporting requirement identified by the project team, you suggest that the project team finalize the formats of all reports and that you will present them to the executives of New National. The project team agrees. The meeting is concluded.

Executive Presentation

When the report formats are finalized, you present the proposed reporting system at an executive meeting. Ms. Karpe and the vice-presidents are impressed with the proposed system. They identify a special ad hoc management report they need to monitor special activity. Ms. Karpe indicates that she and the vice-presidents want an on-line daily report, listing customers in the following categories (HINT: This is a good application for a decision table):

1. Any customer whose checking account balance is $20,000 or more.
2. Any customer whose checking account balance is $10,000 or more who does not have a savings account.
3. Any customer who wrote a check for $10,000 or more.
4. Any customer who had $20,000 or more in savings during the day.
5. Any customer who deposited more than $30,000 in savings or withdrew more than $20,000.
6. Any customer with more than $10,000 in savings who does not own a home.
7. Any customer who has an installment loan with more than $10,000 balance.
8. Any customer who has a commercial loan with more than $500,000 balance.
9. Any customer who has a mortgage loan with more than $150,000 balance.

Ms. Karpe explains that the bank wants to know or become familiar with customers in these categories. She says that the customer master display format would be satisfactory for the report. She wants to access these accounts alphabetically by customer name and feels that a coding system should be set up to explain why a customer is listed on the report. For example, a code value of 1 following the customer information could mean a customer with over $100,000 in savings. Obviously, more than one code value might be applicable to a customer at any point in time. You agree to design such a coding system.

After this discussion, Ms. Karpe directs you to complete the overall systems specifications for the new system. The meeting is then concluded.

Requirements to Complete Case

Complete or design all outputs from, and inputs to, the system; create a data dictionary; construct any necessary decision tables and decision trees; identify the necessary files and construct a data model; develop a menu, and construct a DFD. Complete as much as you can of the design specifications after each meeting.

In some instances, the exact input or output format and validation rules are not told here. You are at liberty to define such issues on a judgmental basis.

There are many ways this case can be expanded into a major, time-consuming effort. To avoid this, stay within the requirements of the case.

The best way to begin is to take the first report, complete the output layout, and make initial entries in the data dictionary for all data elements in that output. Do the same with the second output, the third, and so forth. If you have a CASE tool at your disposal, the output layout can be linked directly to the data dictionary. This will also help ensure consistency in your data definitions.

Chapter 4

Memorial Hospital

Hospitals are called upon to provide an increasing variety of patient services. Medical, surgical, chemical, biological, therapeutic, dietary, housekeeping, and transportation techniques are exemplary of the complex array of services offered. To be effective, hospitals are having to increase the speed and personalization with which they respond to demands for hospital resources.

As with all cases in this book, this case is structured to be realistic and representative of the industry for which it is designed. The case is abbreviated so that it can be completed in a short period of time. In particular, the number of reports and terminal displays is reduced. So is the number of data elements. However, the case does include commonly recognized hospital reporting. ■

Introduction

You have been hired recently as a systems analyst by Memorial Hospital. The chief administrators of Memorial have decided to develop a new information system for the hospital. You have been assigned as the systems analyst to participate in the project. Mr. Dock, hospital administrator, and Dr. Cornette, chief of staff, have asked you to attend an executive meeting of the hospital to discuss the new information system.

Executive Meeting

At the executive meeting you are presented an abbreviated version of Memorial's organization chart (see Figure 4-1). During the meeting it is explained that the hospital is divided into two major organizational groups. The physicians, headed by Dr. Cornette, are ultimately responsible for their patients. The group headed by Mr. Dock provides the clinical, nursing, and administrative support required by the physicians to service their patients.

Figure 4-1 Organization Chart for Memorial Hospital

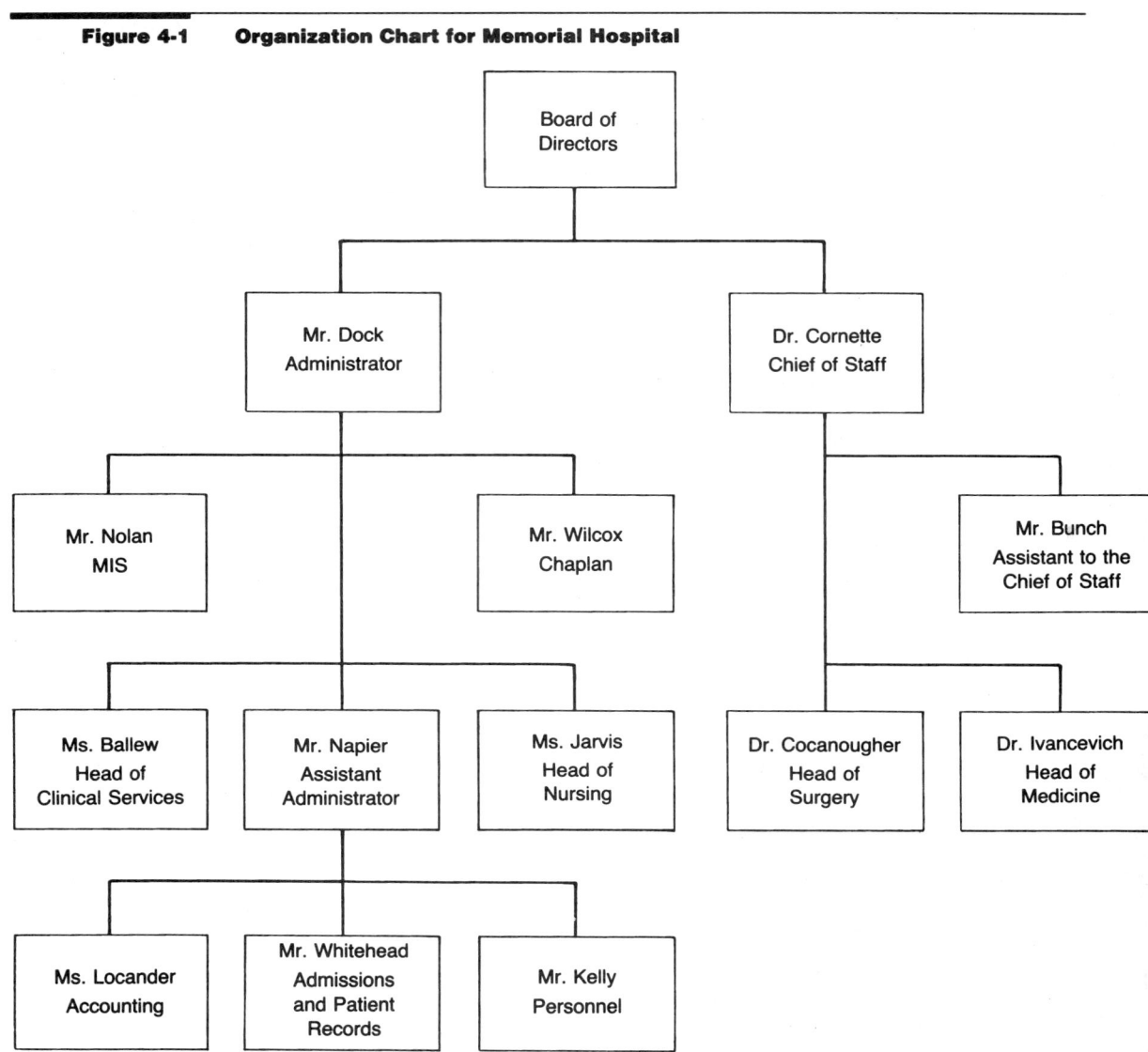

Mr. Dock explains that operational management of the many services needed for a large population of patients requires coordinated reporting from Memorial's information systems. Also, individual services must be accounted for in detail. Each patient is entitled to an explanation of each charge he or she incurs. Prompt financial action from third-party coverage (e.g., insurance companies) is dependent on detailed statements showing the proper division of costs between patients and third parties.

Dr. Cornette explains that Memorial needs to consider not only the schedules and inventory replenishments that are a part of twenty-four-hour operation but also the reallocaton of existing resources and plans for those that are not yet available. Accordingly, the hospital's information system must tally the frequency of demand for various services and print cost summaries for departments providing services.

Mr. Dock adds that financing the long-term capital required for hospital expansion and improvement is most successfully approached when the community is supportive of the hospital's efforts. An understanding community attitude is fostered by regular and specialized explanations of Memorial's contribution to the community. Statistical summaries of hospital services provided can play a key role in keeping the community informed.

Mr. Dock and Dr. Cornette further explain that they want a new information system that addresses hospital and patient reporting. They indicate that, due to the critical nature of the accuracy and timeliness of information, they favor using on-line terminals technology.

You suggest that a project team be organized to conduct the necessary analysis and design the new information system. You request that the project team be composed of high-level staff representatives from all major hospital functions affected by the new system.

Both Mr. Dock and Dr. Cornette concur with your suggestion. The following individuals are assigned to the project team:

- Mr. Dock, hospital administrator and project sponsor
- Yourself, systems analyst and project coordinator
- Mr. Bunch, assistant to the chief of staff
- Ms. Ballew, head of clinical services
- Mr. Napier, assistant administrator
- Ms. Jarvis, head of nursing

Each team member is charged with determining the information requirements of his or her organizational area. You are assigned overall responsibility for coordinating the team efforts and for developing the systems specifications.

(Note: the requirements for outputs from the system should be derived using heuristic or prototyping techniques as described in Chapter 10 of *Systems Analysis and Design: Best Practices*. Since this is not practical to do in a case study, however, outputs are provided for you.)

First Project Team Meeting

You organize a meeting of the project team to initiate the project. You ask each team member to determine the reporting requirements of his or her area of the organization, explaining that once these have been determined, you will be able to "back into" the inputs and processing needed to generate the reports. You also explain that there is cost associated with generating computer-based information, so only information necessary for operations and decision making should be included in the reports.

You offer to coordinate the work of team members to ensure integration of the information reporting requirements. For example, patient data are required by the responsible physician, various clinics, the admissions department, accounting department, nursing department, and so on. Accordingly, each area needs convenient access to patient data without inefficient duplication of record keeping.

Second Project Team Meeting

After extensive analysis of the information requirements at Memorial Hospital, a second project team meeting is called. At this meeting, the project team begins consolidating the various information requirements into several reports.

Daily and Monthly Revenue Report

Mr. Napier discusses the need for an on-line daily revenue report of all dollar transactions. He provides a sample report as a suggestion for the report format (see Figure 4-2).

You ask Mr. Napier whether PATIENT NUMBER is assigned by the hospital and whether it is always five digits long. He responds that PATIENT NUMBER is assigned by the admissions office and that it is always five digits long.

Figure 4-2   Daily Revenue Report

DATE 05-31-97		MEMORIAL HOSPITAL DAILY REVENUE REPORT				PAGE 1
PATIENT NUMBER	LOCATION	FINANCIAL STATUS	COST CENTER	ITEM DESCRIPTION	AMOUNT	TOTAL
54321	3021-1	BLUE-CR	112	ROOM—PRIVATE	80.00	
	3021-1	BLUE-CR	125	PHARMACY—SOLUTIONS	15.00	
	3021-1	SELF-PAY	112	TELEVISION	4.00	
	3021-1	BLUE-CR	132	X-RAY—CHEST	16.00	
						115.00
71623	2125-2	MEDICARE	112	ROOM—4 RM SUITE	40.00	
	2125-2	MEDICARE	101	PATHOLOGY	26.50	
	2125-2	MEDICARE	025	PHARMACY—PENICILIN	14.80	
						81.30

TOTAL PATIENTS	TOTAL CHARGES
620	$36,732.50

He explains the contents of the remainder of the report as follows:

1. DATE—day for which the report is printed. Its format is mm-dd-yy.
2. PAGE—page number of the report. It is initialized to 1 and incremented by 1 for each page.
3. LOCATION—five-digit field (with a hyphen preceding the last digit) indicating the room and bed a patient is assigned to. The first digit indicates the floor where the room is (there are five floors). The next three digits indicate the room (there are up to 125 rooms on each floor). The last digit indicates which bed in the room a patient is in.
4. FINANCIAL STATUS—source of payment for hospital services. This information is coded on patient records to save space; therefore, table lookups are to be used in printing the report. The possible code values and their meanings are:
   1 = BLUE CROSS
   2 = OTHER-BC (a company other than Blue Cross)
   3 = MEDICARE
   4 = MEDICAID
   5 = COMM-INS (commercial insurance)
   6 = WORK-COMP (workmen's compensation)
   7 = WELFARE
   8 = SELF-PAY
   9 = OTHER
5. COST CENTER—three-digit code that uniquely identifies a cost center (pharmacy, x-ray, rooms, etc.). The hospital currently has seventy-five cost centers, identified by the code values from 101 to 175. The name of a cost center can be looked up in a list provided by the assistant administrator. (Only the code values are required on this report.)
6. ITEM DESCRIPTION—the description of the service provided by the cost center. It can be up to twenty-five characters long.
7. AMOUNT—the dollar charge for an item.
8. TOTAL—the total day's charges for a patient. It is computed by adding the AMOUNT values for the patient.
9. TOTAL PATIENTS—the total number of patients having transactions during the day. TOTAL PATIENTS is printed on the last page of the report.
10. TOTAL CHARGES—the total of all charges during the day (i.e., the sum of the TOTAL column). It is printed on the last page of the report.

You ask Mr. Napier where ITEM DESCRIPTIONS and AMOUNT come from. He explains that, as patients receive services during the day, the cost centers fill out forms indicating patient names and numbers, cost-center codes, and services rendered. These forms go to the accounting department during the day. A charge is assigned for each service in accordance with the hospital's pricing manual.

You suggest that for the new system the transactions be entered on-line. This appraoch is agreeable to the project team.

Mr. Napier adds that the hospital needs on-line access to a monthly version of the daily revenue report. You indicate this would be no problem to provide. The daily transactions can be stored and a report generated at the end of the month. (Note that this second report requires a different heading.) Mr. Napier comments that both the daily and monthly versions of the output should be accessible in patient-number sequence.

Room Control Report

Ms. Jarvis and Mr. Napier discuss the need for an on-line daily room control report. They explain that such a report would facilitate bed planning and staffing and provide the statistical basis for facility and room expansion. Mr. Napier points out that the admissions office desperately needs accurate information on admissions, discharges, and transfers in order to control room assignments.

Ms. Jarvis and Mr. Napier provide a sample format for the room control report (see Figure 4-3). Some of the contents of the report are already defined in the previous report. They define the new items as follows:

1. ACCOM—the room accommodations. Possible code values and their meanings are:
   PR = PRIVATE
   SP = SEMI-PRIVATE
   3R = 3-BED WARD
   4R = 4-BED WARD
   IC = INTENSIVE CARE
2. PATIENT NAME—up to thirty characters, printed as follows: LAST NAME, FIRST NAME, MIDDLE NAME OR INITIAL.
3. SEX—coded as follows:
   M = MALE
   F = FEMALE

Figure 4-3   Room Control Report

DATE 5-31-97		ROOM CONTROL REPORT					PAGE 1
ROOM NUMBER	ACCOM	PATIENT NAME	NUMBER	SEX	AGE	ADMIT DATE	EXPECTED DISCHARGE
1001-1	PR	JOHNSON, GEORGE	58723	M	38Y	05-28-97	06-05-97
1002-1	PR	LITZSINGER, JUDY	46115	F	31Y	05-20-97	06-10-97
1003-1	SP	FRANCISCO, AL	78711	M	22Y	05-31-97	
1004-2	SP	IDOL, CHARLES	43244	M	28Y	05-29-97	06-10-97

4. AGE—patient's age. When the age is followed by Y, the age is in years. When the age is followed by M, the age is in months.
5. ADMIT DATE—date the patient was last admitted.
6. EXPECTED DISCHARGE—date the patient is expected to be discharged. If this is not known, the field is blank.

The report is to be accessed room-number order. For example, if the room number 1001-1 is keyed in, it will appear first on the screen, followed by subsequent room numbers.

Scheduled Admissions Report

Mr. Napier states that the admissions office needs an on-line daily scheduled admissions report so that office personnel can prepare for admissions. Mr. Napier provides a sample report format (see Figure 4-4). Much of the information contained in the report is already defined. Mr. Napier defines the new items as follows:

1. DOCTOR—doctor assigned to a patient. This is supplied to the admissions office when a reservation is made.
2. EXPC DATE—date a patient is expected to be admitted. The format is mm-dd.
3. TELEPHONE—phone number where the patient can be reached. It consists of the area code followed by the phone number.
4. THIRD PARTY—primary third party responsible for financial payment. It can be up to twelve characters long.

Figure 4-4 Scheduled Admissions Report

DATE 05-31-97

MEMORIAL HOSPITAL
SCHEDULED ADMISSIONS REPORT

PAGE 1

PATIENT NAME	PATIENT NUMBER	DOCTOR	EXPC DATE	ACCOM	AGE	SEX	TELEPHONE	THIRD PARTY
ADAMSON, RAY	46781	WERNER, M.A.	06-01	PR	16Y	M	713-461-7821	BLUE CROSS
MARR, NANCY	37842	PENDER, R.P.	06-01	SP	32Y	F	713-444-3681	AETNA
PRICE, JOHN H.	17784	CHILDS, J.	06-01	PR	11M	M	612-336-8791	NCR GROUP
WILCOX, HARRY	33614	PENDER, R.P.	05-01	4R	45Y	M	713-456-8217	MEDICARE

Mr. Napier says the report is to be sequenced by PATIENT NAME (alphabetically) within EXPC DATE and should be accessible by patient name or number.

Physician-Patient Report

Mr. Bunch discusses the need for a daily physician-patient report to be sure that all physicians have an accurate account of all their patients currently in the hospital. He provides a sample format for the report (see Figure 4-5).

Mr. Bunch points out that much of the report contents are already defined in previous reports. The new information is defined as follows:

1. EXT—hospital phone extension of a patient. It is four digits long and always begins with a 3.
2. STREET ADDRESS—home street address of a patient. It can be up to fifteen characters long.
3. CITY-STATE-ZIP—home city, state, and zip code of a patient. It can be up to thirty characters long.

Mr. Bunch requests that besides being on-line, the report also be printed on paper in LOCATION sequence to assist physicians in making their daily rounds. On-line access should be based upon doctors' names.

After Mr. Bunch's request, the project team meeting is concluded. A third project team meeting is scheduled to discuss patient records information and billing.

Figure 4-5   Physician-Patient Report

		MEMORIAL HOSPITAL PHYSICIAN—PATIENT REPORT		
DATE 05-31-97		PHYSICIAN—WERNER, M.A.		**PAGE 1**
LOCATION	EXT	PATIENT NAME	STREET ADDRESS	CITY—STATE—ZIP
1025-2	3162	ESTES, NAOMI	1146 PONDEROSA	HOUSTON, TEXAS 77116
1026-1	3175	JENKINS, MILT	1146 OLD OAKS	HOUSTON, TEXAS 71384
1030-3	3160	WILLIAMSON, MARY	1816 ATASCOCIDA	PHOENIX, ARIZONA 16871
1040-1	3180	DEERE, PEGGY	1106 POCATELLO	HOUSTON, TEXAS 78113
1050-1	3184	GREBE, DICK	1207 MOUNTAIN VIEW	ALBUQUERQUE, N.M. 55445

Third Project Team Meeting

Patient Master Display

At the third project team meeting, the first issue discussed is a terminal display to be used for input and retrieval of patient information. After much deliberation, the project team agrees that the following information should be contained in the terminal display. (The information that is not contained in previous displays or is not self-explanatory is defined.)

1. PATIENT NUMBER
2. PATIENT NAME
3. STREET ADDRESS
4. CITY-STATE-ZIP
5. TELEPHONE
6. RESPONSIBLE PARTY—person to be contacted in case of an emergency. This is a twenty-five-character field; last name precedes first name.
7. RESPONSIBLE PARTY'S STREET ADDRESS
8. RESPONSIBLE PARTY'S CITY-STATE-ZIP
9. RESPONSIBLE PARTY'S TELEPHONE
10. SOCIAL SECURITY NUMBER
11. SEX
12. MARITAL STATUS
13. BLOOD TYPE—one of the following values:
    A + or A −
    B + or B −
    AB + or AB −
    O + or O −
14. BIRTH DATE—formatted mm-dd-yy.
15. ALLERGY CODES—two digit code values, separated by commas. There can be up to five code values.
16. RELIGION CODE—one-digit code value, ranging from 1 to 5. Values are defined as follows:
    1 = Protestant
    2 = Catholic
    3 = Jewish
    4 = Other
    5 = Not applicable
17. LOCATION—used only when patient is in the hospital.
18. EXPECTED DISCHARGE—used only when patient is in the hospital.
19. EXPC DATE
20. ADMIT DATE
21. DOCTOR
22. DISCHARGE DATE—date a patient was last discharged. It is used only when applicable.

23. THIRD PARTY
24. FINANCIAL STATUS
25. EXT
26. ADMIT DIAGNOSIS DESCRIPTION—twenty-character field for documenting doctor's preliminary diagnosis.
27. ADMIT DIAGNOSIS CODE—standardized six-digit code that can be cross-referenced to a diagnosis code book.
28. FINAL DIAGNOSIS DESCRIPTION—twenty-character field for documenting doctor's final diagnosis.
29. FINAL DIAGNOSIS CODE—same as Number 27 in this list.
30. FINAL DIAGNOSIS DATE
31. DATE OF DEATH—used only whe applicable.

The project team asks you to design formats for the patient master display. You point out that, depending on the terminal screen size and the format used, two or more "pages" (screen layouts) may be required to display all of the information. This is acceptable to the project team.

You ask whether patient information exists for individuals not currently in the hospital. Mr. Napier indicates it does. Any individual admitted to the hospital becomes a permanent part of the hospital's records. You suggest it may be wise to move records that have been inactive (i.e., records for patients who have not been re-admitted for a prolonged period of time) from the on-line storage to off-line storage. Mr. Napier likes this idea. A ten-year inactive period is established as the criterion for removing patient records from the on-line file.

Next you ask what keys are appropriate for retrieving patient records. The project team agrees that they want to be able to access patient records by both patient name and patient number.

Chaplain's Report

Mr. Napier requests an on-line report for the chaplain. Mr. Wilcox has developed a sample report format (see Figure 4-6). Such a report would allow him to more effectively call on and comfort patients and would also help him to coordinate bringing in a minister of a patient's own faith if desired. Mr. Napier indicates that the report should be in LOCATION order within RELIGION CODE. A new report heading should be generated for each religion code.

Patient Bill

Mr. Napier discusses the billing document (see Figure 4-7). He explains that the information on the bill is familiar to all, with the following exceptions:

1. CODE—same as COST CENTER (see Figure 4-2). The word CODE is used for reference purposes since the cost-center concept is not familiar to most patients. One addition to the previously defined code values is 200, which is used to identify patient payments. Entered payments are subtracted from PATIENT SHARE.
2. CHARGES—same as AMOUNT (see Figure 4-2).

Figure 4-6   Chaplain's Report

```
              MEMORIAL HOSPITAL
              CHAPLAIN'S REPORT
               RELIGION CODE 1

   LOCATION       PATIENT NAME        AGE

    1001-1        PRICE, FRANCIS      25Y

    1001-2        PENDER, RONNY       35Y

    1002-1        ADAMSON, NANCY      18Y

    1003-1        ORR, CHARLES        65Y
```

3. THIRD-PARTY COVERAGE—the portion of the charges covered by a third party. This figure is determined and entered by the accounting department.
4. PATIENT SHARE—the difference computed by subtracting THIRD PARTY COVERAGE from CHARGES.
5. BALANCE DUE—the total PATIENT SHARE.

Mr. Bunch asks if the billing information should be provided as an additional terminal display. Mr. Napier agrees that it would be helpful to the accounting department if an additional display were developed for verifying bills. He points out that charges and payments could be entered on-line. You agree to design a terminal display for the billing information.

Since this is the final reporting requirement identified by the project team, you suggest that the project team finalize the formats of all reports and that you will present them to the chief administrators of the hospital. The project team agrees. The meeting is concluded.

Executive Presentation

When the report formats are finalized, you present the proposed reporting system at an executive meeting of the chief administrators of the hospital. Mr. Dock and Dr. Cornette are impressed with the proposed system. They identify three additional reports required by management.

First, Mr. Dock says he needs a monthly income-analysis report showing dollar volume and transaction volume by cost center. He further wants the report

Figure 4-7   Patient Bill

```
                    MEMORIAL HOSPITAL
            1157 10TH STREET    HOUSTON, TEXAS
                      713-455-6781

     STATEMENT OF ACCOUNT FOR:

         GLASS, CHERYL                                06-25-97
         1218 CLEARVIEW
         HOUSTON, TEXAS 77073           ADMIT        05-15-97
                                        DISCHARGED   05-16-97
                                        PATIENT
                                        NUMBER       54321
```

CODE	ITEM DESCRIPTION	CHARGES	THIRD PARTY COVERAGE	PATIENT SHARE
112	ROOM PRIVATE	80.00	80.00	.00
125	PHARMACY–SOLUTION	25.00	20.00	5.00
112	TELEVISION	4.00	.00	4.00
132	X-RAY–CHEST	16.00	16.00	.00
200	PAYMENT			5.00–
	TOTAL	125.00	116.00	4.00

BALANCE DUE $4.00

to identify the dollar volume by the source of payment. After discussing the report contents, you and Mr. Dock are able to construct a report content (see Figure 4-8). Mr. Dock explains that MONTHLY TRANSACTIONS are the total number of ITEM DESCRIPTIONS entered by a COST CENTER (see Figure 4-2). DISTRIBUTION is based on the FINANCIAL STATUS code values of the patients receiving services from the COST CENTER.

Dr. Cornette discusses the need for a monthly frequency distribution of FINAL DIAGNOSIS CODE (see "Patient Master Display"). The report need only indicate each possible diagnosis code value and a count of the patients classified in each. You agree to design a screen layout.

Mr. Dock and Dr. Cornette say they want an exception report enabling them to monitor patient discharges. The report is to show DATE, PATIENT NAME, PATIENT NUMBER, LOCATION, ADMIT DATE, DOCTOR, ADMIT DIAGNOSIS CODE, and AGE. This information is to be printed for a patient under the following conditions (HINT: This is a good application for a decision table):

1. If the patient's ADMIT DIAGNOSIS CODE (ADC) is between 000000 and 000501, hospital stay is more than three days and age is less than sixty-five years *or* hospital stay is more than two days and age is sixty-five or older.

Figure 4-8 Monthly Income Analysis

DATE 06-30-97

MEMORIAL HOSPITAL
MONTHLY INCOME ANALYSIS

PAGE 1

COST CENTER	MONTHLY TRANSACTIONS	MONTHLY CHARGES	BLUE CROSS	OTHER-BC	DISTRIBUTION MEDICARE	MEDICAID	DISTRIBUTION COMM INS	WORK COMP	WELFARE	SELF-PAY	OTHER
101	522	18,746.00	10,702.00	2,040.00	568.00	1,120.00	2,561.00	785.00		970.00	
102	68	90,842.75	46,832.25	33,010.50			2,862.00			3,138.00	
103 -	782	64,781.32	58,651.38	5,480.30						649.64	

2. If ADC is between 000500 and 001001, hospital stay is more than three days, and age is less than sixty-five *or* hospital stay is more than five days and age is sixty-five or greater.
3. If ADC is between 001000 and 006001, hospital stay is more than five days, and age is less than sixty-five *or* hospital stay is more than eight days and age is sixty-five or greater.
4. The patient is currently in intensive care (a message should be printed indicating this).

The report is to be accessible in patient-number sequence within ADC sequence. You are told to design a screen layout.

After discussing the last report, Mr. Dock directs you to complete the overall systems specifications for the new system. The meeting is then concluded.

Requirements to Complete Case

Complete or design all outputs from, and inputs to, the system; create a data dictionary; construct any necessary decision tables and decision trees; identify the necessary files and construct a data model; develop a menu, and construct a DFD. Complete as much as you can of the design specifications after each meeting.

In some instances, the exact input or output format and validation rules are not told here. You are at liberty to define such issues on a judgmental basis.

There are many ways this case can be expanded into a major, time-consuming effort. To avoid this, stay within the requirements of the case.

The best way to begin is to take the first report, complete the output layout and make initial entries into the data dictionary for all data elements in that output. Do the same with second output, the third, and so forth. If you have a CASE tool at your disposal, the output layout can be linked directly to the data dictionary. This will also help ensure consistency in your data definitions.

Chapter 5

Intermountain Distributing Inc.

Distributing companies are engaged primarily in the purchasing of finished goods inventory from manufacturing and/or other distributing companies. Accordingly, the primary function of a distributing company is to provide "place" and "time" utility for merchandise by locating and storing merchandise in locations convenient for future sales.

As with all cases in the book, this case is structured to be realistic and representative of the industry for which it is designed. The case is abbreviated so that it can be completed in a short period of time. In particular, the number of reports and terminal displays is reduced. So is the number of data elements. However, the case does include the most important and commonly recognized reporting in the distribution industry. ∎

Introduction

You have been hired recently as a systems analyst for Intermountain Distributing Inc. The chief executive officers of Intermountain have decided to develop a new information system for the company. You have been assigned as the systems analyst to participate in the project. Mr. Grebe, president of Intermountain, has asked you to attend an executive meeting with the vice-presidents to discuss the new information system.

Executive Meeting

At the executive meeting you are presented a copy of Intermountain's organization chart (see Figure 5-1). During the meeting, it is explained to you that Intermountain's operations are based on the following major functions:

- *Sales and customer service:* Mr. Wismer, vice-president of marketing, is responsible for this function. It consists of managing sales personnel, advertising,

Figure 5-1 Organization Chart for Intermountain Distributing Inc.

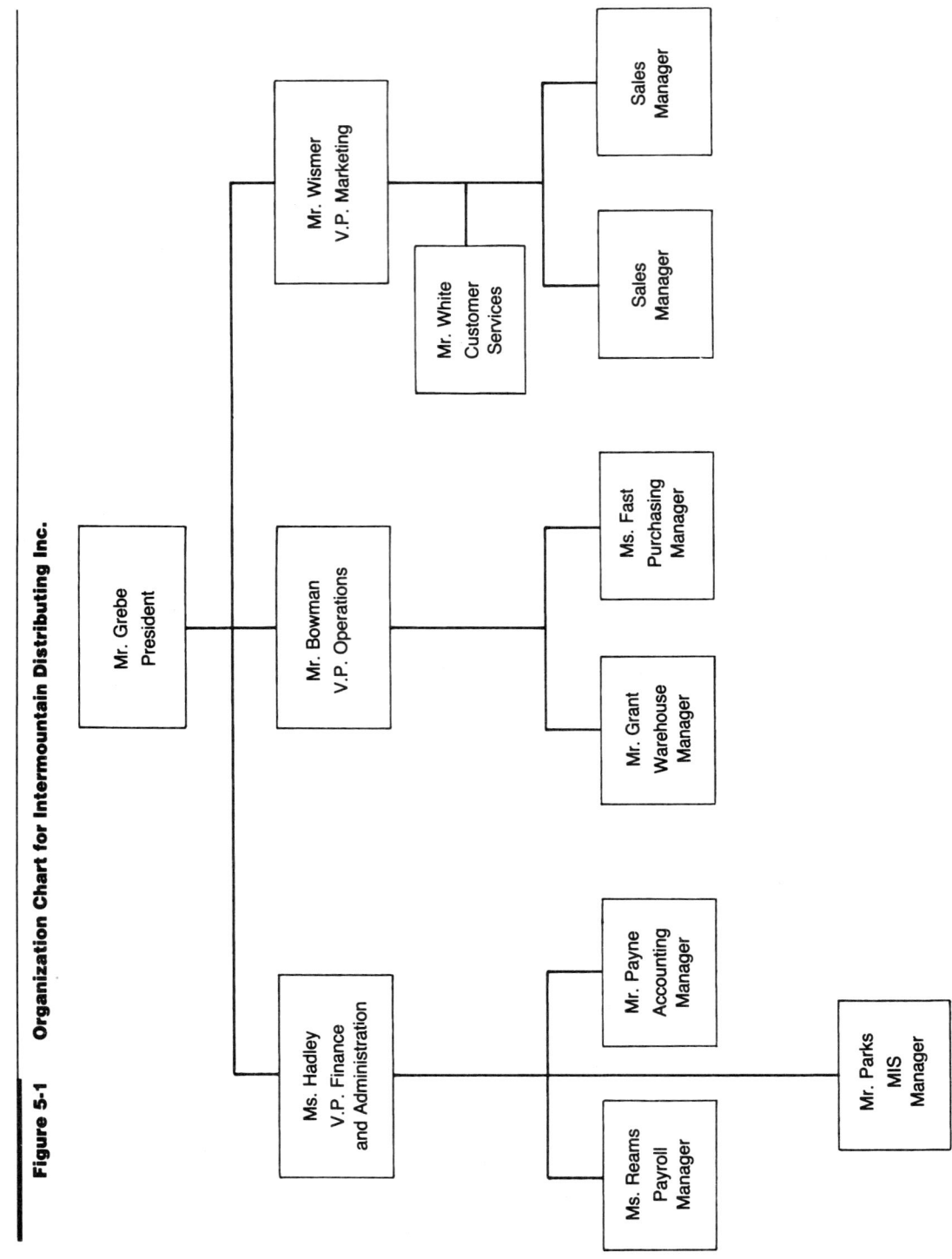

and sales promotion; establishing and executing customer service procedures; and processing sales to the operations division.

- *Operations management:* Mr. Bowman, vice-president of operations, is responsible for this function. It consists of processing orders from the sales department; order filling and shipping; and inventory management and purchasing.

Financial and administrative management: Ms. Hadley, vice-president of finance and administration, is responsible for this function. It consists of general ledger, accounts receivable, accounts payable, invoicing, payroll/personnel, and the MIS Department.

Mr. Wismer indicates that it is particularly important to Intermountain to have current, accurate information on the status of orders and inventory. Therefore, the new information system should support on-line processing of customer orders and the effects they have on inventory status. Ms. Hadley also expresses her desire that on-line processing be used to expedite transaction processing and to eliminate the need for a large keypunch staff.

You suggest that a project team be organized to conduct the necessary analysis and design the new information system. You request that the project team be composed of high-level staff representatives from all major functions affected by the new system.

Mr. Grebe concurs with your suggestion. The following individuals are assigned to the project team:

- Mr. Wismer, vice-president of marketing and project sponsor
- Youself, systems analyst and project coordinator
- Mr. Payne, accounting manager
- Mr. Grant, warehouse manager
- Ms. Fast, purchasing manager
- Mr. White, customer services

Each team member is charged with determining the information requirements of his or her organizational division. You are assigned overall responsibility for coordinating the team efforts and for developing the systems specifications.

(Note: The requirements for outputs from the system should be derived using heuristic or prototyping techniques as described in Chapter 10 of *Systems Analysis and Design: Best Practices*. Since this is not practical to do in a case study, however, outputs are provided for you.)

First Project Team Meeting

You organize a meeting of the project team to initiate the project. You ask each team member to determine the reporting requirements of his or her division, explaining that once these have been determined, you will be able to "back into" the inputs and processing needed to generate the reports. You also explain that there is cost associated with generating computer-based information, so only information necessary for operations and decision making should be included in the reports.

You indicate that you will coordinate the team efforts to ensure integration of the information-reporting requirements. For example, inventory status is used by the purchasing department for reordering; by the warehouse to determine shipping; by the sales department to determine delivery; and by the accounting department to value inventory for the general ledger. Accordingly, each department needs to have convenient access to accurate information on inventory status.

Second Project Team Meeting

After extensive analysis of the information requirements at Intermountain, you call a second project team meeting. At this meeting, the project team consolidates the various information requirements into several reports.

Inventory Management

Ms. Fast, purchasing manager, indicates that the purchasing and customer-services departments require an on-line exception display indicating all inventory items that are out of stock or below their minimum reorder points. Such a report would warn of low stock conditions, facilitate control of stock outages, and allow the customer-services department to notify customers so that substitutions or other alternatives can be considered. Ms. Fast provides a sample format as a suggestion for the report (see Figure 5-2).

You ask Ms. Fast whether VENDOR NUMBER and ITEM NUMBER are always numeric and whether they are always six and eight digits in length, respectively. Ms. Fast said they are unless an error is made. She also informs you that ITEM DESCRIPTION is up to fifteen characters long, and that REORDER POINT and QUAN ON ORD (quantity on order) are never greater than 999 and 9999, respectively. STOCK CONDITION is to indicate either OUT or BELOW MIN (minimum) depending on the value of ON HAND. Whenever ON HAND equals 0 (zero), four asterisks are to be printed below the ON HAND heading. DATE ORDERED is the last date an order was placed—format is MM-DD-YY.

You ask Ms. Fast if the report should be in any sequence. She responds that it should be sequenced according to item number.

Ms. Fast further tells you that the credit and marketing functions need current information on each inventory item. She proposes that this requirement be accomplished through terminal displays of the entire contents of any inventory record.

All project team members favor this idea. You suggest that this same display could be used for data entry (e.g., additions, deletions, and modifications) to inventory. It is decided, therefore, that on-line capabilities will be incorporated for update and inquiries of inventory data. You agree to design the format for the terminal display.

Vendor Reporting

Mr. Grant mentions the need for on-line data entry and retrieval of data about vendors who supply inventory to Intermountain. Ms. Fast agrees and further

Figure 5-2 Out of Stock/Below Minimum Report

			OUT OF STOCK/				
DATE 12-06-97			BELOW MINIMUM REPORT				PAGE 1
VENDOR NUMBER	ITEM NUMBER	ITEM DESCRIPTION	STOCK CONDITION	ON HAND	REORDER POINT	QUAN ON ORD	DATE ORDERED
146894	67894220	HAIR DRYER	OUT	****	60	250	12-02-97
792267	68116772	ELECTRIC SHAVER	BELOW MIN	20	40	100	12-03-97
406843	46214558	GUITAR	OUT	****	50	175	12-01-97
787743	22116600	LAMP	BELOW MIN	10	40		12-04-97
432687	98987146	CHESS SET	BELOW MIN	42	100	300	12-01-97

indicates that the only information that the purchasing department needs about a vendor is vendor number (a six-digit field), name, address, city, state, zip code, telephone number, and the minimum dollar order the vendor is willing to ship. You agree to prepare a screen format for the listing of vendor data. Access to vendor data will be based upon vendor number or vendor name.

Invoicing

Mr. Payne, accounting manager, and Mr. White, of customer services, next propose the contents and format for customer invoices, as illustrated in Figure 5-3. Mr. Payne explains that both a billing address (i.e., SOLD TO) and a shipping address (i.e., SHIP TO) are required. By asking additional questions, you determine the definition of other data elements as follows:

1. DATE—date the invoice is created.
2. PAGE—number of this page of the invoice. It is necessary when more than one page is required to list all items ordered.
3. ORDER NO.—six-digit number assigned to the order when it is entered into the system from a terminal.
4. INVOICE NO.—six-digit number assigned to the invoice as it is printed. For each batch of invoices on a given day, the invoice number is initialized to 1, and it is incremented by 1 for each invoice generated. Thus, given a date and invoice number, a specific invoice can be indentified.

Figure 5-3   Invoice Format

INTERMOUNTAIN DISTRIBUTING, INC.

DATE 12-06-97
PAGE 01

SOLD TO

THE TOY STORE, INC.
300 BROWN ST.
P.O. BOX 8275
DENVER, COLORADO 76142

SHIP TO

THE TOY STORE
LORETTO PLAZA
415 MAIN ST.
PORTLAND, OREGON 68172

ORDER NO.   123456
INVOICE NO.   123456
CUSTOMER NO. 123456

ITEM NUMBER	DESCRIPTION	U/M	ORDERED	QUANTITY SHIPPED	B/O	RETAIL PRICE	NET PRICE	EXTENDED PRICE
68246789	JET ROCKET	DZ	12	12	0	12.95	6.95	83.40
42687420	BIONIC MAN	EA	10	5	5	8.50	4.40	22.00
68714321	SKATE BOARD	BX	24	0	24	10.25	8.75	.00
77882169	DOLL HOUSE	EA	1	1	0	22.95	11.45	11.45
30002468	GUITAR	EA	6	6	0	165.00	95.50	510.00

TAX	SHIPPING	QUANTITY DISCOUNT	INVOICE TOTAL	RETAIL TOTAL	NET TOTAL
91.33	53.20	275.60	1,918.00	3,848.00	2,049.07

5. CUSTOMER NO.—six-digit number assigned to each customer by the accounting department.

6. U/M—unit of measure for an item. Options are:
   EA = Each
   BX = Box
   DZ = Dozen
   FT = Foot
   YD = Yard

7. QUANTITY ORDERED, SHIPPED, and B/O (Backordered)—three four-digit fields. Backorders apply to items not currently in stock that have to be shipped at a later date.

8. RETAIL PRICE—suggested retail price for each item. It should never exceed $2,000.00.

9. NET PRICE—price to the buyer. It should never exceed $2,000.00.

10. EXTENDED PRICE—price for the total number of units shipped (i.e., QUANTITY SHIPPED × NET PRICE). It should never exceed $10,000.00.
11. NET TOTAL—total of the EXTENDED PRICE column.
12. RETAIL TOTAL—total of RETAIL PRICE × QUANTITY SHIPPED for each item.
13. QUANTITY DISCOUNT—total quantity discount for the order. It is computed as follows: Each inventory item has up to four quantity break levels with the following discounts:
Quantity Level A = 1.0%
Quantity Level B = 1.5%
Quantity Level C = 2.0%
Quantity Level D = 3.0%
The specific numeric value assigned to a quantity level is determined by the vice-president of marketing and can vary among inventory items. For example, quantity break levels may be assigned to two inventory items as follows:

Quantity Level	Guitars	Guitar Picks
A	5	100
B	10	250
C	15	500
D	20	1000

During invoice preparation, the computer must compare the quantity ordered with the quantity break-level values to determine the percent to use for computing the quantity discount for each item ordered. The total discount for all items on an invoice determines the QUANTITY DISCOUNT for the invoice (HINT: This is a good application for a decision table):

14. SHIPPING = 3% of (NET TOTAL − QUANTITY DISCOUNT).
15. TAX = 5% of (NET TOTAL − QUANTITY DISCOUNT + SHIPPING).
16. INVOICE TOTAL = NET TOTAL − QUANTITY DISCOUNT + SHIPPING + TAX.

You ask if quantity-level data (see QUANTITY DISCOUNT) should be added to the terminal display for inventory items. Ms. Fast says that is an excellent idea and to please do so.

Though invoices will be printed on forms, an on-line version copy of the invoice needs to be available to respond to customer inquiries. The on-line version needs to be accessible by date plus either customer number or name.

Mr. Payne says that Ms. Hadley has requested an on-line daily customer transaction report to monitor sales activity. The report should contain one line for each invoice and report the invoice number, customer number, customer name, retail total, net total, quantity discount, tax, shipping charge, and invoice total. The information would need to be accessed by date, plus either customer number or name. An appropriate heading should be provided. Mr. Payne asks you to design the screen format.

With Mr. Payne's request, the project team meeting is concluded. A third project team meeting is scheduled to discuss warehouse and credit requirements.

Third Project Team Meeting

Filling and Shipping Orders

At the third project team meeting, Mr. Grant discusses the requirements for filling and shipping orders. He indicates that a picking list and a packing slip are required for processing orders in the warehouse. The picking list, he explains, is used to determine the items ordered by a customer and the bin locations in the warehouse where the items are stored. He proposes the format illustrated in Figure 5-4 for the picking list.

You ask Mr. Grant to explain the format of the BIN LOCATION. He responds that the first character indicates the warehouse and it can be coded as an A, B, or C. The warehouse code is followed by a blank and four digits separated by a dash. The first two digits indicate the aisle number, which ranges from 01 to 50. The last two digits indicate the bin numbers within an aisle. Bin numbers range from 01 to 99.

The packing slip is enclosed in the package for shipping. It is used by the customer to validate the goods received. The format that Mr. Grant proposes for the packing slip is illustrated in Figure 5-5. Mr. Grant explains that WEIGHT is the weight of the U/M (unit of measure).

After analyzing the picking list and the packing slip, you notice they are quite similar in content. You raise the question of consolidating the contents in one document. This could be accomplished by adding the bin location to the packing slip.

Figure 5-4   **Picking List**

DATE 12-08-97		PICKING LIST			PAGE 01
CUSTOMER NO.		CUSTOMER NAME		ORDER NO.	
123456		THE HANDY HUT		123456	
BIN LOCATION	ITEM NUMBER	ITEM DESCRIPTION	U/M	SHIP	WEIGHT
A 01-13	46872431	LIPSTICK	BX	5	1.2
A 01-17	61127421	HAIR SPRAY	BX	10	4.0
A 03-04	77864230	HAIR DRYER	EA	20	24.6
B 07-06	46789106	SIDE MOLDING	FT	100	.5
•		•	•	•	•
•	•		•	•	•
	•	•			

TOTAL WEIGHT 121.6

Figure 5-5 Packing Slip

DATE 12-08-97		PACKING SLIP			PAGE 01
		CUSTOMER NO. 123456			
SOLD TO		SHIP TO		ORDER NO.	123456
				ORDER DATE	12-01-97
THE HANDY HUT		THE HANDY HUT			
CENTER BUILDING		CENTER BUILDING			
P.O. BOX 8172		P.O. BOX 8172			
ATLANTA, GEORGIA 71682		ATLANTA, GEORGIA 71682			

ITEM NUMBER	ITEM DESCRIPTION	U/M	SHIPPED	WEIGHT	SUGGESTED RETAIL
46872431	LIPSTICK	BX	5	1.2	1.29
61127421	HAIR SPRAY	BX	10	4.0	1.95
77864230	HAIR DRYER	EA	20	24.6	12.75
46789106	SIDE MOLDING	FT	100	.5	8.95

TOTAL WEIGHT 121.6

Mr. Grant asks what advantages this approach offers. He also says that he has heard of multiple-part paper forms and wants to know if these can be used. You answer that by consolidating the contents you can have the two documents printed at the same time. You also suggest that while multiple-parts forms are popular, many organizations are now using high-speed laser printers to print multiple copies of the same form. Mr. Grant tells you to use the printing method that you deem to be appropriate. This would save computer time and expedite the processing of picking and packing information needed at the warehouse. Mr. Grant and the other project team members accept your suggestion. You agree to design a new combined picking list/packing slip document.

Mr. Grant points out that, as with invoices, an on-line version of the picking list/packing slip document needs to be available for internal use at Intermountain. It needs to be accessible by date plus either customer number or order number.

Customer and Credit Reporting

The next topic discussed at the meeting is the customer information needed by the accounting and marketing functions. Mr. Payne explains that some of the required customer data has been identified in previously defined reporting (e.g., customer number, customer name, billing and shipping addresses, and order data). However, the accounts-receivable department has to approve credit for each order processed. Credit approval consists of adding the dollar amount of each incoming order to the customer's current balance. This new balance is then compared to the customer's credit limit. If the new balance is less than the credit limit, the order is approved. Otherwise, the determination of approval or rejection of credit is handled on an exception basis. Each exception usually involves the supervisor of accounts receivable, who in turn contacts Mr. White in Customer Services, and possibly the customer, to decide an appropriate course of action.

You ask Mr. Payne how customer payments are handled. He explains that customer payments are cycled through the accounts receivable department and deducted from customer balances. He comments that it would be nice if the new system allowed accounts receivable personnel to deduct payments from balances via computer terminals. You agree to this approach.

Next, you ask Mr. Payne if it would help the efficiency of the accounts receivable department if the computer automatically processed routine credit authorizations. All exceptions could be reported, via terminals, to the accounts receivable department for special attention.

Mr. Payne is agreeable to this plan. He asks how such a system would operate. You explain that all orders would be entered into the system by order-processing clerks using terminals in their offices. As the orders are entered, the computer system would perform the credit checks. If credit is approved, the order would be admitted to the system. If credit is not approved, the system would notify the order-processing clerk to hold the particular order until further notification from the accounts receivable department. The system would then generate an exception report on a terminal in the accounts receivable department. The report would be used to resolve the credit problem. If the credit is approved, the accounts receivable department could send an electronic mail message to the order-processing department and instruct them to re-enter the order. If credit is not approved, a message could be sent to customer services for final disposition.

Mr. Payne is impressed with this approach. You ask him to define the information required on the exception report. He lists the following items:

1. CUSTOMER NAME
2. CUSTOMER NUMBER
3. CUSTOMER PHONE (area code included)
4. YEAR ACCOUNT OPENED—six-digit field of the form MM-DD-YY.
5. CREDIT LIMIT—not to exceed $999,999.00.
6. CREDIT RATING—one-digit field coded as follows: 1 = Excellent, 2 = Good, 3 = Poor.
7. CREDIT RATING DATE—six-digit field of the form MM-DD-YY.
8. DATE OF LAST SALE—MM-DD-YY.
9. CURRENT BALANCE—not to exceed $999,999.00.
10. CURRENT ORDER AMOUNT—amount of the order that, when added to the current balance, caused the exception report to occur.

You agree to develop a simple format suitable for a terminal to report credit authorization exceptions to the accounts receivable department.

Since this is the final reporting requirement identified by the project team, you suggest that the project team finalize the formats of all reports and that you will present them to the executives of Intermountain. The project team members agree. The meeting is concluded.

Executive Presentation

When the report formats are finalized, you present the proposed reporting system at an executive meeting. Mr. Grebe and the vice-presidents are impressed with the proposed system. They identify two additional reports required by management.

First, Mr. Grebe and Mr. Bowman discuss the need for a summary report indicating the profitability of each inventory item. They say the report should indicate past-month and year-to-date totals for each inventory item. The totals to be reported follow:

- UNITS SOLD—total number of units sold.
- UNIT OF MEASURE—measure used to define unit (e.g., box, dozen, each).
- NUMBER OF INVOICES—total number of invoices prepared.
- GROSS SALES—total dollar value of items sold (i.e., units sold × unit price).
- DISCOUNTS—total of all discounts as computed during invoice generation.
- SALES EXPENSE—computed as follows:
NUMBER OF INVOICES × $36.00 + 15% of GROSS SALES.
- NET PROFIT—GROSS SALES less SALES EXPENSE and DISCOUNTS
- % ITEM PROFIT—NET PROFIT divided by GROSS SALES.

Mr. Bowman explains that the profitability-by-item report should be generated at the end of each month, at which time it could be available for on-line access. Year-to-date totals should be initialized to zero at the end of the calendar year. Past-month totals should be computed each month and added to the year-to-date totals.

Mr. Grebe asks you to design an appropriate format for the report. You agree to do so.

Next, Mr. Wismer and Mr. Grebe present the need for a profitability-by-customer report. They indicate that the report should provide on-line access to year-to-date and past-month totals for each customer as follows:

- NUMBER OF INVOICES
- GROSS SALES
- DISCOUNTS
- SALES EXPENSE
- NET PROFIT
- % CUSTOMER PROFIT—NET PROFIT divided by GROSS SALES.

The initialization and monthly adjustments to totals are identical to those for the profitability-by-item report. You agree to design a report format and make sure that it is accessible by customer number and name.

After discussing the two management reports, Mr. Bowman directs you to complete the overall systems specifications for the new system. The meeting is then concluded.

Requirements to Complete Case

Complete or design all outputs from, and inputs to, the system; create a data-element dictionary; construct any necessary decision tables and decision trees; identify the necessary files construct a data model; develop a menu, and construct a DFD. Complete as much as you can of the design specifications after each meeting.

In some instances, the exact input or output format and validation rules are not specified here. You are at liberty to define such issues on a judgmental basis.

There are many ways this case can be expanded into a major, time-consuming effort. To avoid this, stay within the requirements defined in the case.

The best way to begin is to take the first report, complete the output layout, and make initial entries into the data dictionary for all data elements in that report. Do the same with the second report, the third, and so forth. If you have a CASE tool at your disposal, the output layout can be linked directly to the data dictionary. This will also help ensure consistency in your data definitions.

Chapter 6

Medical Supply Inc.

Medical Supply Inc., is engaged primarily in manufacture and distribution of medical supplies and equipment. Medical Supply installed an order processing system (OPS) 3 years ago. Although the system met the expectations of the original users, the heart valve division, it is recognized that many new businesses and organizations have begun using newer systems and that the existing system may not fully meet training needs now or in the future. ∎

Introduction

You have been hired as a systems analyst for Medical Supply Inc. The chief executive officers of Medical Supply have decided to develop a new order processing system for the company. You have been assigned as the systems analyst to participate in the project. Mr. Knapp, president of Medical Supply Inc., has asked you to attend an executive meeting with the vice presidents to discuss the new information system.

Executive Meeting

At the executive meeting you are presented with a copy of Medical Supply's organization chart (see Figure 6-1). You are asked to describe your impression of the useful life expectancy of information systems. You suggest that it is widely recognized that most business application systems have a useful life of no more than 6 or 7 years. During that time the business requirements and/or the available technology changes in significant enough ways as to render a system obsolete. Although it can continue to be used, usually the expense of maintaining the system begins to place financial burdens on the user organization.

Based on your comments, it is decided to begin the process of finding a suitable replacement for OPS, recognizing that the search and subsequent installation might take 3 years. By that time, OPS would be over 6 years old.

Figure 6-1   Organization Chart for Medical Supply Inc.

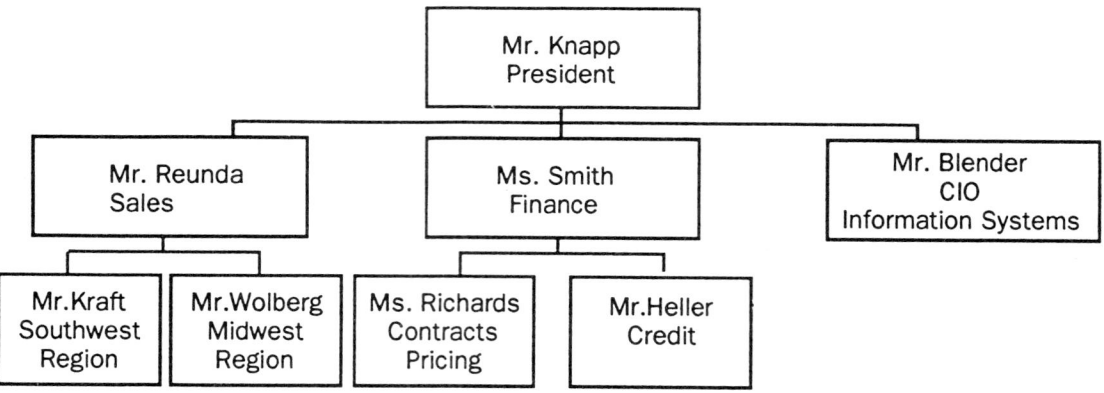

A study team of users, IS personnel, and business managers is then assembled consisting of the following members:

- Bob Blender, IS, Chairperson
- Yourself, Systems Analyst and Project Coordinator
- Elaine Smith, Finance
- Faye Richards, Contracts
- Steve Heller, Credit
- Avery Tollerman, Physical Distribution
- Bob Kraft, Southwest Regional
- Tracy Wolberg, Midwest Regional

Each team member is charged with determining the information requirements of his or her organizational division. You are assigned overall responsibility for coordinating the team efforts.

First Project Team Meeting

You organize a meeting of the project team to initiate the project. The overall mission of your project is broken into three phases:

1. Phase 1: Examine the future technology and business requirements and determine if OPS should be replaced with a new purchased system, replaced with a Medical Supply–built system, or remodeled.
2. Phase 2: Select the proper replacement package or tools to complete the project.
3. Phase 3: Program, modify, test, and install the new OPS system.

Only Phase 1 is included in the scope of the project at this time. At the end of this task, the team will be reevaluated to determine if different members are needed for Phases 2 and 3.

Once it is determined that Phase 1 is the only phase to be addressed at this time, you expose the team to the concept of *vision-based strategic planning*. This form of planning methodology was developed by CSC/Index. Vision-based planning involves visualizing how you want the business to look and how you want to do business at some time in the future. This is different from attempting to predict the future in that vision-based planning implies that you have control to make the future happen as you see it. The vision created then implies some actions that have to be taken now to make it happen or some interim stages that must occur to make certain you are on track toward reaching your vision.

Figure 6-2 depicts the approach used during Phase 1. The team was divided into two groups, one for technology and one for business. The technology group, through library research, seminars and reading, and consultations with other IS experts, developed a vision of what IS technology was likely to look like in 4 or 5 years. This approach, because Medical Supply has no ability to affect the development of IS technology, is truly an attempt to predict the future. It is based on the premise that any technology commercially available in 4 to 6 years is demonstrable in the lab today.

The business group used a more classic vision-based planning technique. Based on information contained in the company's long-range plan documents, statements made by business unit leaders, and the personal knowledge and expertise of group members, a vision of Medical Supply's business characteristics

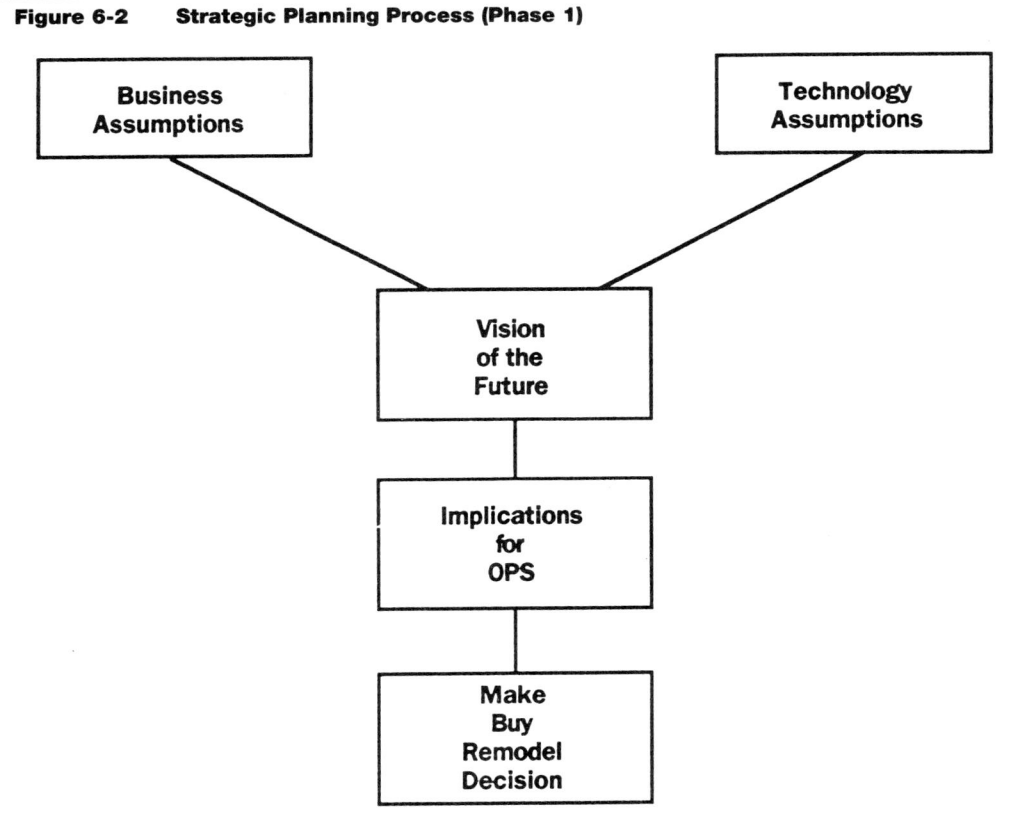

Figure 6-2   Strategic Planning Process (Phase 1)

was developed. As a check against the vision of many of Medical Supply's top managers, a survey was developed to address the key vision elements. The vision was adjusted based on the survey returns.

With both visions in hand, the implications to the OPS system were developed and a final strategy recommended for its replacement.

You inform the team that based on the vision-based strategic planning agenda (see Figure 6-2) they should focus on (1) a technology-based vision of the future, (2) a business-based vision of the future, and (3) the implication of 1 and 2 on OPS.

Second Project Team Meeting

After extensive analysis of the future of technology, a second project team meeting is called. At this meeting, the project team consolidates what they believe to be the future of technology.

The visions of the future in the 5-year timeframe for computer technology are broken into three main categories: (1) Workstations, (2) Telecommunications and Networks, and (3) Midrange Computers and Software.

Workstation Assumptions

1. Computer power will continue to increase during the next 5 years and will equal the tenfold increase experienced during the last 5 years.
2. With this additional power, windowing, multitask processing, and color will become the norm.
3. Friendly user interfaces using mice, bar code readers, touch-sensitive screens, and pen interfaces will continue to grow in popularity and will become the norm.
4. Multimedia systems utilizing text, images, motion video, and sound will begin to appear but will likely still be on the leading edge.
5. *User-friendliness* will become the watchword. Users will not be required to understand the operating system of a workstation. All functions will be intuitive and controlled by graphical user interfaces or pens. Functions, commands, and the general look and feel of the workstation will be highly customizable by the individual user to suit individual preferences.
6. Cooperative processing will increasingly become a common practice. As defined here, cooperative processing involves a business system whose processes are shared by several computers (servers) and workstations. The amount of processing that occurs at the central processor and at the workstation varies by application. Where central access to data and higher processing power are needed, a central processor may be used. Where fast response and local intelligence are advisable, the workstation takes over. Another term frequently associated with co-processing is the *client-server model*.

Telecommunications and Networks Assumptions

1. Wide area networks 5 years from now will not be oriented toward leased, hardwired lines as they are predominantly today but will be software-controlled public data networks where transmission of data between two points will be charged by the unit. These networks will detect failure within the network and reroute the traffic without the user's application being affected. As a result, the applications design will assume that long distance communications are fast and reliable.

2. Wide area network speeds between Kansas City and a regional office will increase from 56kb to 384kb during the next 5 years at the same cost, allowing the practical use of cooperative processing.

3. International network speeds will increase from 56kb in today's environment to 128kb in 5 years at the same cost. Cooperative processing will be less of an opportunity here than locally in the United States.

4. ISDN (Integrated Services Digital Network) will not become a viable tool during the planning period, but will require at least 10 years to be widely available.

5. Token Ring will eventually win the battle of the local area network protocols over its competitors. It will become more reliable, full featured, and faster.

6. Society and work structures will continue to change as a result of continually improving communications. In the next 5 years, we will begin to see some employees working from home using workstations and high-speed communications, but it will not be widespread.

7. The lines between a company and its vendors and customers will continue to blur in the next 5 years. Electronic connections for order entry, data query, funds transfer, and the general exchange of information will continue to grow.

8. SDNs (Software Defined Networks) will change the possibilities with respect to serving the customer on an around-the-clock basis. This telecommunications technology will enable organizations to be designed that can serve the customer from central locations and still provide the personal service found in today's regional locations.

Midrange Computers and Software Assumptions

1. As with the workstation, midrange computers will continue to grow in power, tenfold in the next 5 years.

2. Hardware costs, particularly storage, will continue to drop while increasing in speed. More data will be kept on-line, immediately accessible to the users. Database managers will evolve at the same time, allowing transparent distributed databases to become a reality.

3. User-friendliness will improve as computer power allows changes to operating systems. AT (Artificial Intelligence) elements will begin to appear in systems. With wide ranges of information available to the individual user and AI available to help find and interpret that information, cross-functional systems will become more prevalent. A single employee will have the information available to act as a single contact point for customers.

4. Image processing (the computer storage, manipulation, and display of pictures) will become one of the most important factors in computer technology for the 1990s.
5. The concept of a *knowledge farm* that contains vast amounts of information on many subjects for the company will continue to expand, allowing more users to access the information.
6. *Open systems* will continue to grow in popularity and proprietary operating systems will diminish. That is not to say that UNIX operating systems will displace current proprietary operating systems such as OS400 at IBM or VMS at DEC, but rather that all operating systems will slowly evolve and grow together to standardization.
7. Fourth-generation languages may be replaced in the next 5 years by fifth-generation, utilizing simple English and AI. The result will be greater flexibility and maintainability in systems. As with midrange hardware, we may be seeing systems in the next 5 years that never will have to be replaced but rather will evolve over the years to meet the changing needs of the business.
8. Packages will continue to be used widely for traditional business applications. However, such new techniques as cooperative processing and graphical user interfaces will be slow to appear in packages. Those businesses who wish to stay on the leading edge of the technology will have to consider in-house development.
9. *Object orientation* at the presentation layer and later at the application logic layer will continue to evolve at an accelerating pace.
10. Outsourcing of IS services will continue to grow. IS professionals will tend to provide consulting services to users as experts in applying information technology to real business problems. Outside services or the users themselves will provide the implementation. Outsourcing vendors will continue to evolve a range of services in the marketplace.

Summary

The advancement in computing power and storage will enable future systems to incorporate more corporate information and power at the user's desk. Cooperative processing with transparent, graphical, and user-friendly interfaces will prevail.

Advanced languages, object orientation, and artificial intelligence will change the relationship between IS professionals and the users. IS will provide the computing power, communications, and consulting. The users and outside vendors will provide more of the actual system development.

Telecommunications and networking will improve, providing the possibility of an environment in which the concept of distributed databases and cooperative processing over distances will be transparent to the user.

In order to stay competitive and to meet the quality objectives, systems in the future must provide Medical Supply employees with more information in a simpler fashion.

At the conclusion of the second project meeting, you congratulate the team on conceptualizing the future of technology. You ask the team to focus their attention for the next meeting on the future vision of the business.

Third Project Team Meeting

After 3 weeks, the project team meets again to discuss the future vision of Medical Supply. In developing a vision of where Medical Supply wishes its business to be 5 years hence and some of the projected influences on that vision, 11 separate categories were examined.

Distribution Channels

1. No major changes in distribution channels will occur in the next 5 years, with one exception. The increase in drug sales will require a whole new set of channels and requirements that current systems are not equipped to service.
2. Current customers (i.e., hospitals) will continue to constrict their suppliers. Consolidation among suppliers will continue. Competition to be one of those fewer vendors will become greater, requiring better customer service as a competitive tool.
3. Although direct selling will remain in the channel of choice, in the international environment, distributors will continue to be used where they make sense. Increased use is not expected.

Inventory

1. Volume customers will not stock at their own expense. A limited number of customers will hold small amounts of inventory. There will be greater pressure to provide fast response to the user's product needs.
2. Consignment inventory and trunk stock will play a more important role in Medical Supply's inventory strategy 5 years hence. Control and visibility into that inventory by all concerned will become more important.
3. Distribution center stocking will become decentralized by product. One single warehouse for all products will be eliminated. The need will be to order, ship, and invoice from multiple locations in a manner that is transparent to the customer.

Accounting

1. Multiple legal entities will remain a reality both in the United States and internationally.
2. Innovative financing arrangements such as service contracts, leasing, and rentals will grow in importance.
3. Installment purchases will grow and become more creative and flexible.
4. There will be increased attention to capital accounting and amortization.
5. Cost accounting within the distribution organization will become more important. In addition, cost accounting methods will change in the next 3 to 5 years, although their exact form is not known at this time.

Pricing

1. Bundling (selling product as a kit consisting of a sensor and leads, for example) will become important in many business units because of the convenience afforded to the customer. The important characteristic here is that the kits will not stock, only the individual components. The "kitting" takes place when the product is shipped. The price is determined at the kit level.
2. Contract volume will increase.
3. Promotional pricing, beyond contracts—such as quantity buys and special, limited-time offers—will become important.
4. Terms will become important as a marketing tool and an alternative to promotional pricing.
5. Volume rebates will continue.

Credit

1. Total account management will become a reality. One representative will be responsible for credit, order processing, and other account servicing.

Electronic Data Interchange (EDI)

1. Electronic order entry directly by customers will play a role, especially for commodity products, but will not become dominant.
2. *Electronic funds transfer* will become a reality in the next 5 years.
3. The electronic transmission of sales history information will grow during the planning period.
4. A National Medical Registry, that Medical Supply will feed, will become a reality, requiring EDI applications for medical device registration data.

Product

1. Traceability requirements will increase during the next 5 years. Some features to expect will be physician data required at time of shipment for drugs and patient notification on field actions.
2. Look for all implantable accessories to be serialized in the future.

Business Unit Organization

1. There will be little change in business unit organization during the planning period.
2. Business units will become more independent over time, in an effort to reduce bureaucracy.
3. There will be fewer independent sales forces directly under the control of the business units, although they still will be specialized by product line.

4. Worldwide common order processing systems will not be required. Integration of product, sales, and inventory data will be done at a higher level in the systems.

Sales Organization

1. Sales representatives will evolve toward becoming business managers, supported by specialists. Account managers will support the representative and the accounts with "one stop" access to information and administrative problem solving. And regional inventory specialists will manage inventory across many accounts.
3. No change is anticipated in the sales compensation system beyond the usual annual adjustments currently being experienced.

Physical Distribution Organization

1. The mission and form of the regional offices will change. They will become regional headquarters for management and training facilities only. No inventory will be carried and no customer service activities will take place.
2. Account servicing and order entry will be done at two central customer service locations. The extensive use of SDN technology will enable the organization to maintain the personal relationships that currently exist between customer service representative and customer while at the same time providing wider time coverage. Twenty-four-hour coverage may become a reality.

Changing Customer Needs

1. More common ownership will characterize the medical provider customer base.
2. Customers will rely on suppliers to manage inventory for them.
3. Large customers will rely more on direct channels to order product.

Summary

The basic structure of the business will not change radically in the next 5 years, but the way we conduct that business will. Quality efforts will demand that customer service provide quicker response and more centralized services while still providing the personal touch. Around-the-clock availability will become necessary.

Cost pressures in the medical marketplace, particularly with health care reform, will require more creative pricing and financing options for the customer and will require more services, such as administrative, to be provided by Medical Supply.

Inventory management will be an even more decentralized and important aspect of the business than it is today. Getting the product closer to the customer (from an order-time-to-delivery point of view) will become more important.

After the team discussed the visions of both the future of the technology and the business, implications were derived for presentation to the executives of Medical Supply.

Executive Presentation

The visions of the future both for technology and for the business were related to today's OPS system. Mr. Knapp and the vice presidents were impressed by the thoroughness of the analysis. The implications of the future of technology and the business on the OPS were summarized as follows:

1. The responsiveness of the system today falls short in being able to provide quick entry and query of information while dealing with a customer on the phone. In the future, multiple groups of information will be required on the user's screen simultaneously in order to service the account adequately.

2. The operating window (7 A.M. to 8 P.M. central time) will not be sufficient in the future. The goal will be 24-hour availability for data entry, not just inquiry.

3. Requirements for batch reporting as we know it today for inventory analysis, sales commission reporting, sales reporting, and the like will remain basically unchanged. Some new reports will be required and some existing reports may be obsolete.

4. Separate systems for Europe and the United States and perhaps other international geographic regions will continue to satisfy the company's needs, as long as the core systems remain the same and data is integratable worldwide.

5. The inventory allocation and shipping capability within the existing OPS will not meet the grade in tomorrow's environment of decentralized inventory. This OPS has problems in ordering, shipping, and invoicing products shipped from multiple locations.

6. Some of the new channel and market requirements from business units cannot be met by the OPS today. The drug market will strain the system, and future blanket orders with periodic releases will cause the users to develop manual workaround procedures.

7. Pricing, payment terms, and financing options as visualized for the future cannot be completely met with the logic in today's system.

8. Kitting is not possible in today's OPS.

9. More vigorous inventory management capability will be required as consignment, trunk, and decentralized inventory become more prevalent. The ability to manage inventory across accounts will become a necessity.

10. Cross-functional requirements demanded by tomorrow's physical distribution and credit organizations cannot be met by today's system. Only one piece of information can be displayed to the user at a time. Such features as Automatic Caller Identification, allowing basic information about the caller to be displayed automatically by the system, are not now possible.

11. Today's OPS is designed for central processing. In its current form it cannot take advantage of new technology in the form of cooperative processing and workstations as they become available.
12. The current system cannot handle EDI applications. Significant modifications and new programs will be required to couple this OPS with EOE and EFT applications.

Summary

It is believed that the basic data structure of the existing OPS, although cumbersome in some cases and in need of simplification and cleanup, provides the capabilities needed by the organization well into the future. Batch reporting may change slightly, but the foundation is there. The users have spent a great deal of time and effort in getting it to its present state.

The functionality of the OPS that falls short of meeting tomorrow's needs is centered in the logic for order entry, shipping, invoicing, pricing, and inventory management.

In addition, the existing system is unable to react to changing technology and the reengineering of the organization that is made possible by that technology. Centralizing customer service organizations, decentralizing inventory, and improving CFQ will require good cross-functional systems and user-friendliness. The OPS, in its current form, is stuck in the past. It is serving today's needs but cannot react effectively to the future.

The executives of Medical Supply were concerned by the possible future of the OPS. Your team has been charged with moving to Phase 2 of the project.

Requirements to Complete Case

To complete the case and satisfy the requirements of Phase 2, please consult the tables on pages 81 to 88. Also note that some of the required capabilities, illustrated in the tables and needed by Medical Supply, do not exist as current capabilities.

1. Flowchart the OPS process.
2. Analyze information requirements in view of present capabilities and required capabilities and redesign the new process. For example, currently cost information is delivered via phone call. A new requirement is to deliver price information via electronic data interchange (EDI) against a price database. New hardware and software will be required.
3. Do a feasibility analysis of the proposed solutions.
4. Do a data flow diagram (DFD) for the new system.
5. Do input/output design, data dictionary, and presentation layer interaction.
6. If the platform is to be client-server, go back to Chapter 4 of the text and identify the requirements and design for each component and layer.

7. Prepare a final presentation for management. This presentation should include the following:
   - Executive Summary
   - Introduction
   - Purpose of Study
   - Method of Analysis
   - Redesigned Solutions
   - Justification
   - Appendices

Table 6-1   Order Entry Process: Customer Access and Maintenance Information

Business Function	Present Capabilities	Required Capabilities

Customer information needs to be accessed for order input. The access and selection process needs to be direct and simple:
  One number input
  Call path options
  Direct key: ph#/st/div
	Many screens, many questions asked of customer	• One # on screen, input—give the correct customer information required
		• Selection ability for multi ship-to/bill-to on one screen

Customer information needs to be entered in the outer locations and transferred electronically to Mpls for review and finalization	Phone call to Kansas City	• Electronic transfer • Auto D & B access • Tax automation
Access to the multi files of customer information needs to be available and accessible in numeric, alpha, combos, zip, phone, group search, etc.	Limited/*no* access to files	• Interrogate the various customer files in many ways
Customer-specific information needs to be stored and available to order entry personnel, i.e., blanket PO, people names, personal information, hospital-specific information, always ship-to—, always bill to—, contract spec., etc.	Limited/none Customer text	• Secured storage/access to personal customer information • Access in order entry
Send verification of customer information to each customer annually to keep file correct and updated	N/A	• System creates an account profile and letter to obtain updated customer information
Customer information should be accessed by most used, i.e., a region should get region accounts in first search but have ability to search deeper information and notification	Selection process includes all customer accounts	• First-level customer information should be more specific to caller or business with additional levels of search
Consignment information/ contract	N/A	• Customer service review/use in servicing the account replenishment responsibilities, inventory management
A means of managing multi-accounts under one umbrella or corporate account structure	N/A	• A need exists to establish unique accounts for division responsibility but with a method of looking at all accounts together as a corporate account
Maintain file by removing old records and transferring information to new record	N/A	• Delete transfer capabilities

Table 6-2   Order Entry Process: Order Types

Business Function	Present Capabilities	Required Capabilities
Regular orders: Process orders that ship product to customer	Manually log Enter later	• Enter during call completion • Process/quote/pick/ship immediately
Back orders: Process orders that hold product for customers	Manually log Enter as regular Default to back order	• Enter as a back order • System manage and ship based on date required/promised
Consignment: Process orders that consign and track inventory in a hospital	Same as regular order	• Same as regular, plus special hospital management capabilities
Consignment/replacement: Process orders that manage the removal and replacement of inventory in the hospital	N/A	• Option at order entry to replace same/different products on hospital consignment without re-keying of same fields
Kit/bundle orders: Process orders that manage the components in packaged products for Regulatory Affairs	Ltd kit Bundle N/A	• Combine finished good products for kit • Ship from multiple warehouses • Ship manufactured component products • Trace components by individual item
Warehouse transfers: Process orders that move/track product to valid inventory locations and accumulate replacement requirements for group shipment	Same as regular order	• Accumulate transfer orders for efficient shipping • Scan to validate serial numbers, quantity, etc.
Warehouse tran/replenish: Process orders that replace sold products automatically that move/trace product to valid inventory locations and accumulate replacement product requirements for group shipment each day/week/month, etc.	N/A	• Dual purpose order to effectively transfer/replenish product in warehouse stock—see consignment order for similar requirement
Intercompany orders are not included in this process but are a part of the OPS process for order entry	N/A	N/A

Continues on next page

Table 6-2   (Cont.)

Business Function	Present Capabilities	Required Capabilities
Edit orders on an exception basis and keep manual intervention to a minimum: entry, shipment	Edit customer information Edit product information Edit warehouse information	• Obtain valid customer information with no more than 2 screens • System assigns warehouse for shipment—note only exceptions • Pricing appears automatically—no special request
Pending order entry of an order needs free form capabilities so the order is retained in any partial state for update/continuance until completion is realized	N/A	• Once an order has begun, it should remain on file until complete/canceled • Notify of outstanding orders
Electronic order entry: Phone order entry, fax order entry, voice-activated order entry deserve some attention to create the flexibility needed to handle future volume/productivity	N/A	• Multi technologies are required to satisfy the diverse customer population—we need to help the customer profit so we can be profitable • Share system information electronically
Multi division entry with system handling	OK	• Multi division entry with ability to identify and specify division/association required
Fast for customer service and customer Easy for customer service and customer Flexible for customer service and customer	5 minutes	• 1 minute entry to completion for shipment Limited interrogation of customer • Key one transaction for order entry
Use of bar code reader/scanner in order shipment process to reduce keying effort wherever possible	N/A	• Scan products for shipment and record serial number on order
Inclusion of pertinent link information, i.e., surgery date, patient name, specific association products, doctor name, etc.	Text	This information needs to be stored and utilized for specific purpose: • Surgery date, ship short-dated product • Patient name, hospital requirements
One screen entry	Minimum of 4	• One screen entry required because of speed of customer service

Continues on next page

Table 6-2 (Cont.)

Business Function	Present Capabilities	Required Capabilities
Inquiry access to customer account, customer sales history, inventory, product, contract, plus other related information. Multi order access during order entry.	N/A	• Hot keys (pop-ups) required to inquire into specific areas of information to respond to customer inquiries as well as to assist the customer throughout the order

Handling of drug distribution
  Physician
  Pharmacy name and number
  Patient
  Prescription number
	N/A	• Drug distribution is totally new to the arena; some of the requirements might be:
		  Patient name, address
		  Prescribing physician
		  Prescription number
		  Term of use

Customer-specific editing is required	N/A	• Customer-specific requirements • Purchase order edit • Blanket purchase order • Customer-specific products
Handle blanket POs with special extension field	N/A	• Manage blanket POs with shipping times to completion as well as a PO suffix extension • Plan blanket orders in billing schedules
Third-party payer invoicing	N/A	• Ability to generate/track invoice to third-party payers • Ability to satisfy government regulations • Ability to send electronic claim forms
Electronic orders	N/A	• Ability to receive/process and send pre-ship notification for EDI orders

Lease orders	OK	• Ability to manage:
		  —Various lease types
		  —Various terms
		  —Various ledger entries
		  —Various invoicing req.
		  —Various inventory trans

Blanket orders	N/A	• Electronic or keyed order to create shipments and invoices over a scheduled period of time • System should manage and notify according to established requirements

Table 6-3   Order Entry Process: Inventory Allocation

Business Function	Present Capabilities	Required Capabilities
System allocate inventory by total order	N/A	• System finds and allocates inventory required thru search method established by set criteria
System allocate inventory by logical ship-from location	By regional requirements Default customer file Manual decision	• System finds and allocates inventory required thru search method established by set criteria
Review inventory allocation on-line	Matrix allocation system	• Retain
Allocate inventory by serial number	Post bill Regular order with serial keyed	• Retain • Retain
Manage product shortages	Matrix system managed allocation	• System back order automated • Notify of instances

Provide product information to the customer service person:
    Compatibility
    Availability
    Status
    Product substitutions
	Stock status
	Stock status
	Stock status
	N/A
		• Need information present on the AS/400 on-line when speaking to a customer

Use of bar code readers to reduce keying effort in all areas feasible	Portable and stationary readers All products bar coded	• Assign and verify lots/serial numbers for shipment • Handheld units for cycle counting in the consignment accounts
Easy help access in process pop-ups (field validation) field text	N/A	• Field help • Field allowed codes • Background text

Inventory inquiry
    National
    Product group
    Warehouse
    Bin
    Consignment
    Etc.
	Distribution inquiry
	Bin inquiry
	Bin inquiry
	Bin inquiry
	Bin inquiry
	Bin inquiry
		• Greater selectivity
		• Partial serial/lot
		• Model/Loc/Scan/Point/Click

Perpetual inventory control	N/A	• Interactive updates • Use with bar code readers • Greater automation in updates and write-offs
Safety stock management information	Manage by SKU system	• Greater visibility/information
Short-dated inventory lookup and allocation	N/A	• Allocate when "Rush" date is a fit • Highlight products available with short dates to determine if use is available within order

Table 6-4 **Order Entry Process: Pricing**

Business Function	Present Capabilities	Required Capabilities

All orders should be priced accurately
   Contract prices
   Promotional prices
   Short-term pricing
 	
   Freight
   Tax
   Price override
	Specific contract terms not all accurate—most are
	
	N/A
	
	Old calc—not correct
	Cumbersome
	OK
		100% accuracy
		
		• Temporary pricing/verbal agreements
		• Interactive with shipper
		• Tax-based system, i.e., zip?
		• Retain

All price quotes should be accurate and include freight, tax, and price breaks	N/A	• Quotes/inquiries and orders should price correct and include helpful customer information
Special pricing of grouped products, i.e., when a magnet is given free with the implant purchase, it is free and the order entry process should *not* require an override	Override	• Price automatically—group pricing
Pricing overrides are required in those instances where a one-time special price is given and authorizations are granted	Override	Retain

Price information is required by the customer service representative when in contact with the customer
   Contract expiration dates
   Contract status-pending?
   Price breaks
   Special marketing promotions
	N/A
	Complicated
		• Incorporate inquiries (pop-ups) in the order entry and the price quote processes

• Selection of price information by least costly by customer and product group • Other price/product interrogations	N/A	• Provide multi views of contract information • View customer pricing by expiration date • View contract document(s) by customer or contract
Contract pricing for kitted products by kit and by customer is required	N/A	• Contract on groups of products purchased together from Finished Goods for pricing
System authorization/notify process for special pricing of override and out of contract	N/A	• Notify systematically the correct person to authorize price and receive authorization via the system
Minimum prices allowed	N/A	• Flag price lower than minimum calculation • Special authorization required via system
Lowest price calculation by customer/product	Available	Retain

Table 6-5   Order Entry Process: Shipping

Business Function	Present Capabilities	Required Capabilities
Customer-specific shipping instructions	Default in customer master plus override by order taker	Utilize more economical/effective system method or user choice
Freight calculated based on logical locations, ship method, product and destination	Limited freight calculation ($4.50, $2.50, $1.00, or $.50)	More accurate reflection of actual cost (handling charge?)
Shipment quote information is required when completing a customer quote	Freight charges estimated but represent guess as to services available	Better education or on-line data to quote regarding services and rates
Manage COD process in order entry, shipping, and invoicing	On-line notification	Highlight • Create special shipping documents • Notify billing—hold invoice
Customer Master Files	Limited access Cannot inquire against some files	Multi-file access for different inquiries • Available to access one find for a variety of inquiries
Product Master Files	Available	Inquiry by "wild card" fields • Inquiry to all fields • Inquiry by groups/common name, etc.
Credit Reason Code Master	N/A	Inquire required
Inventory Master Files	Limited available	One inventory inquiry for all information • Select by warehouse/model/lot/expiration date/etc.
Contract Master Files	Contract number needed for inquiry	By customer • Option for unknown contract numbers • By customer group, etc.
Term Master File	N/A	Inquiry to master file showing terms, code, and description
Order Files	Only certain inquiry sequences allowed	Inquiry by many/all sequences
Override Code Master	N/A	Inquiry required
Sales Files —History —Same Day	 Limited available N/A	 Required by multiple searches Required by multiple searches

Table 6-6   Order Entry Process: Help

Business Function	Present Capabilities	Required Capabilities

Field help on screen
    Pop-ups
    Validations
    Background text
	N/A	• By field
		• Valid codes and descriptions
		• By application

Current help information requires updates when changes are made	N/A	• Build into change process and manual updates

Table 6-7   Order Entry Process: Training

Business Function	Present Capabilities	Required Capabilities
On-line tutorial and training sample set	None	• Tutorial system in place
New employees training and certification	No formal training system No certification	• Training system in place • Certification process in place
Employee re-training and re-certification	None	• System in place/annual refresh

DATA-DICTIONARY

NAME:

KEYWORDS:

DEFINITION:

FORMAT:

SOURCE:

VALIDATION RULES:

WHERE USED:

ENTITY:

STORED:

MAINTENANCE:

DATA-DICTIONARY

NAME:

KEYWORDS:

DEFINITION:

FORMAT:

SOURCE:

VALIDATION RULES:

WHERE USED:

ENTITY:

STORED:

MAINTENANCE:

DATA-DICTIONARY

NAME:

KEYWORDS:

DEFINITION:

FORMAT:

SOURCE:

VALIDATION RULES:

WHERE USED:

ENTITY:

STORED:

MAINTENANCE:

DATA-DICTIONARY

NAME:

KEYWORDS:

DEFINITION:

FORMAT:

SOURCE:

VALIDATION RULES:

WHERE USED:

ENTITY:

STORED:

MAINTENANCE:

DATA-DICTIONARY

NAME:

KEYWORDS:

DEFINITION:

FORMAT:

SOURCE:

VALIDATION RULES:

WHERE USED:

ENTITY:

STORED:

MAINTENANCE:

DATA-DICTIONARY

NAME:

KEYWORDS:

DEFINITION:

FORMAT:

SOURCE:

VALIDATION RULES:

WHERE USED:

ENTITY:

STORED:

MAINTENANCE:

DATA-DICTIONARY

NAME:

KEYWORDS:

DEFINITION:

FORMAT:

SOURCE:

VALIDATION RULES:

WHERE USED:

ENTITY:

STORED:

MAINTENANCE:

DATA-DICTIONARY

NAME:

KEYWORDS:

DEFINITION:

FORMAT:

SOURCE:

VALIDATION RULES:

WHERE USED:

ENTITY:

STORED:

MAINTENANCE:

DATA-DICTIONARY

NAME:

KEYWORDS:

DEFINITION:

FORMAT:

SOURCE:

VALIDATION RULES:

WHERE USED:

ENTITY:

STORED:

MAINTENANCE:

DATA-DICTIONARY

NAME:

KEYWORDS:

DEFINITION:

FORMAT:

SOURCE:

VALIDATION RULES:

WHERE USED:

ENTITY:

STORED:

MAINTENANCE:

DATA-DICTIONARY

NAME:

KEYWORDS:

DEFINITION:

FORMAT:

SOURCE:

VALIDATION RULES:

WHERE USED:

ENTITY:

STORED:

MAINTENANCE:

DATA-DICTIONARY

NAME:

KEYWORDS:

DEFINITION:

FORMAT:

SOURCE:

VALIDATION RULES:

WHERE USED:

ENTITY:

STORED:

MAINTENANCE:

DATA-DICTIONARY

NAME:

KEYWORDS:

DEFINITION:

FORMAT:

SOURCE:

VALIDATION RULES:

WHERE USED:

ENTITY:

STORED:

MAINTENANCE:

DATA-DICTIONARY

NAME:

KEYWORDS:

DEFINITION:

FORMAT:

SOURCE:

VALIDATION RULES:

WHERE USED:

ENTITY:

STORED:

MAINTENANCE:

DATA-DICTIONARY

NAME:

KEYWORDS:

DEFINITION:

FORMAT:

SOURCE:

VALIDATION RULES:

WHERE USED:

ENTITY:

STORED:

MAINTENANCE:

DATA-DICTIONARY

NAME:

KEYWORDS:

DEFINITION:

FORMAT:

SOURCE:

VALIDATION RULES:

WHERE USED:

ENTITY:

STORED:

MAINTENANCE:

DATA-DICTIONARY

NAME:

KEYWORDS:

DEFINITION:

FORMAT:

SOURCE:

VALIDATION RULES:

WHERE USED:

ENTITY:

STORED:

MAINTENANCE:

DATA-DICTIONARY

NAME:

KEYWORDS:

DEFINITION:

FORMAT:

SOURCE:

VALIDATION RULES:

WHERE USED:

ENTITY:

STORED:

MAINTENANCE:

DATA-DICTIONARY

NAME:

KEYWORDS:

DEFINITION:

FORMAT:

SOURCE:

VALIDATION RULES:

WHERE USED:

ENTITY:

STORED:

MAINTENANCE:

DATA-DICTIONARY

NAME:

KEYWORDS:

DEFINITION:

FORMAT:

SOURCE:

VALIDATION RULES:

WHERE USED:

ENTITY:

STORED:

MAINTENANCE:

DATA-DICTIONARY

NAME:

KEYWORDS:

DEFINITION:

FORMAT:

SOURCE:

VALIDATION RULES:

WHERE USED:

ENTITY:

STORED:

MAINTENANCE:

DATA-DICTIONARY

NAME:

KEYWORDS:

DEFINITION:

FORMAT:

SOURCE:

VALIDATION RULES:

WHERE USED:

ENTITY:

STORED:

MAINTENANCE:

DATA-DICTIONARY

NAME:

KEYWORDS:

DEFINITION:

FORMAT:

SOURCE:

VALIDATION RULES:

WHERE USED:

ENTITY:

STORED:

MAINTENANCE:

DATA-DICTIONARY

NAME:

KEYWORDS:

DEFINITION:

FORMAT:

SOURCE:

VALIDATION RULES:

WHERE USED:

ENTITY:

STORED:

MAINTENANCE:

DATA-DICTIONARY

NAME:

KEYWORDS:

DEFINITION:

FORMAT:

SOURCE:

VALIDATION RULES:

WHERE USED:

ENTITY:

STORED:

MAINTENANCE:

DATA-DICTIONARY

NAME:

KEYWORDS:

DEFINITION:

FORMAT:

SOURCE:

VALIDATION RULES:

WHERE USED:

ENTITY:

STORED:

MAINTENANCE:

DATA-DICTIONARY

NAME:

KEYWORDS:

DEFINITION:

FORMAT:

SOURCE:

VALIDATION RULES:

WHERE USED:

ENTITY:

STORED:

MAINTENANCE:

DATA-DICTIONARY

NAME:

KEYWORDS:

DEFINITION:

FORMAT:

SOURCE:

VALIDATION RULES:

WHERE USED:

ENTITY:

STORED:

MAINTENANCE:

DATA-DICTIONARY

NAME:

KEYWORDS:

DEFINITION:

FORMAT:

SOURCE:

VALIDATION RULES:

WHERE USED:

ENTITY:

STORED:

MAINTENANCE:

DATA-DICTIONARY

NAME:

KEYWORDS:

DEFINITION:

FORMAT:

SOURCE:

VALIDATION RULES:

WHERE USED:

ENTITY:

STORED:

MAINTENANCE:

DATA-DICTIONARY

NAME:

KEYWORDS:

DEFINITION:

FORMAT:

SOURCE:

VALIDATION RULES:

WHERE USED:

ENTITY:

STORED:

MAINTENANCE:

DATA-DICTIONARY

NAME:

KEYWORDS:

DEFINITION:

FORMAT:

SOURCE:

VALIDATION RULES:

WHERE USED:

ENTITY:

STORED:

MAINTENANCE:

DATA-DICTIONARY

NAME:

KEYWORDS:

DEFINITION:

FORMAT:

SOURCE:

VALIDATION RULES:

WHERE USED:

ENTITY:

STORED:

MAINTENANCE:

DATA-DICTIONARY

NAME:

KEYWORDS:

DEFINITION:

FORMAT:

SOURCE:

VALIDATION RULES:

WHERE USED:

ENTITY:

STORED:

MAINTENANCE:

WARNING! This is the last blank form. Copy this if you need more.

SCREEN LAYOUT

SCREEN LAYOUT

SCREEN LAYOUT

SCREEN LAYOUT

SCREEN LAYOUT

SCREEN LAYOUT

SCREEN LAYOUT

SCREEN LAYOUT

WARNING! This is the last blank form. Copy this if you need more.

REPORT DEFINITION for

_____
Application

REPORT LAYOUT

NAME OF REPORT _____
PREPARED BY _____
DATE _____ PAGE ___ OF ___

REPORT DEFINITION for _____
_____ **Application**

NAME OF REPORT _____
PREPARED BY _____
DATE _____ **PAGE** ____ **OF** ____

REPORT LAYOUT

REPORT DEFINITION for
_____ **Application**

I REPORT LAYOUT

NAME OF REPORT _____
PREPARED BY _____
DATE _____ **PAGE** _____ **OF** _____

REPORT DEFINITION for _____ **Application**

REPORT LAYOUT

NAME OF REPORT _____
PREPARED BY _____
DATE _____ **PAGE** ___ **OF** ___